Introduction to Quantum Mechanics 2

Introduction to Quantum Mechanics 2

Wave-Corpuscle, Quantization & Schrödinger's Equation

Ibrahima Sakho

iSTE

WILEY

First published 2020 in Great Britain and the United States by ISTE Ltd and John Wiley & Sons, Inc.

ISTE Ltd
27-37 St George's Road
London SW19 4EU
UK

www.iste.co.uk

John Wiley & Sons, Inc.
111 River Street
Hoboken, NJ 07030
USA

www.wiley.com

Library of Congress Control Number: 2019950855

British Library Cataloguing-in-Publication Data
A CIP record for this book is available from the British Library
ISBN 978-1-78630-501-5

MIX
Paper from responsible sources
FSC® C013604

Contents

Foreword

Founded in 1925 and 1926 by Werner Heisenberg, Erwin Schrödinger and Paul Dirac, quantum mechanics is nearly 100 years old. As the basis of modern technology, it has given rise to countless applications in physics, chemistry and even biology. The relevant literature is very rich, counting works written in many languages and from various perspectives. They address a broad audience, from beginner students and teachers to expert researchers in the field.

Professor Sakho has chosen the former as the target audience of this book, connecting the quarter of a century that preceded the inception of quantum mechanics and its first results. The book is organized in two volumes. The first deals with thermal radiation and the experimental facts that led to the quantization of matter. The second volume focuses on the Schrödinger equation and its applications, Hermitian operators and Dirac notations.

The clear and detailed presentation of the notions introduced in this book reveals its constant didactic concern. A unique selling point of this book is the broad range of approaches used throughout its chapters:

– the course includes many solved exercises, which complete the presentation in a concrete manner;

– the presentation of experimental devices goes well beyond idealized schematic representations and illustrates the nature of laboratory work;

– more advanced notions (semiconductors, relativistic effects in hydrogen, Lamb shift, etc.) are briefly introduced, always in relation with more fundamental concepts;

– the biographical boxes give the subject a human touch and invite the reader to anchor the development of a theory in its historical context.

The book concludes with a list of references and a detailed index.

Science is a key element of contemporary culture. Researchers' efforts to write the books required for students' education are praiseworthy. Undergraduate students and teachers will find this work especially beneficial. We wish it a wide distribution.

Louis MARCHILDON
Professor Emeritus of Physics
University of Quebec at Trois-Rivières
July 2019

Preface

Quantum mechanics or the physics of the infinitely small (microcosm) is often contrasted with classical mechanics or the physics of macroscopic bodies (macrocosm). This book, whose title is "Introduction to Quantum Mechanics 2", aims to equip the reader with basic tools that are essential for a good understanding of the physical properties of atoms, nuclei, molecules, lasers, solid bodies and electronic materials – in short all that is infinitely small. Introductory courses on quantum mechanics generally focus on the *study of the interaction between matter and radiation, and the quantum states of matter*. This book emphasizes the various experiments that have led to the discovery within the set of physical phenomena related to the properties of quantum systems. Consequently, this book is composed of seven chapters organized in two volumes. Each chapter starts with a presentation of the general objective, followed by a list of specific objectives, and finally by a list of prerequisites essential for a good understanding of the concepts introduced. Furthermore, the introduction of each law follows a simple application. Each studied chapter ends with a collection of various rich exercises and solutions that facilitate the assimilation of all the concepts presented. Moreover, a brief biography of each of the thinkers having contributed to the discovery of the studied physical laws or phenomena is given separately, as the chapter unfolds. The reader can this way acquire a sound scientific culture related to the evolution of scientific thought during the elaboration of quantum mechanics. Due to its structuring and didactic approach, this work is a modern and very original book. Volume 1 covers the study of the first four chapters related to *thermal radiation*, to the *experimental facts* that revealed the *quantization of matter*, and to *De Broglie wave theory and Heisenberg's uncertainty principle*.

Volume 2 is dedicated to the last three chapters related, respectively, to the study of Schrödinger equation and applications, Hermitian operators and Dirac notations.

Chapter 1 focuses on the study of the evolution of wave functions described by the *Schrödinger equation* followed by several applications that introduce, in particular, concepts such as *potential well*, *potential path*, *wave reflection* and *transmission factor*, *potential barrier*, *tunnel effect* and *0D confinement* through the study of *quantum dots*. Chapter 2 deals with the basic tools related to the mathematical formalism of quantum mechanics. Hence, this chapter presents the properties of *orthonormal bases* in the *space of* square-summable *wave functions*, *Dirac notations* for *ket* and *bra vectors* in the *state space*. Moreover, it introduces notions such as *linear operator*, *Hermitian operator*, *observable*, *Hermitian conjugation* and *commutator*. Finally, Chapter 3 studies the eigenvalues and eigenvectors of an observable. This offers the possibility to introduce the notion of *representation* of ket and bra vectors and operators, to pass from *vector calculus* in the space of square-summable wave functions and to *matrix calculus* in the space of states. Furthermore, the study relates to the introduction of the *eigenvalue equation* of an operator and the *characteristic equation* (or *secular equation*) for determining the eigenvalues of an operator based on a matrix representation. The chapter ends with the definition of the *mean* value of an *observable* and the establishment of their *evolution equation* by the study of *conservative systems*, and the establishment of *Ehrenfest theorem* reflected by the laws of evolution of the mean values of *position* and *momentum operators*.

Finally, the book is completed by a set of appendices that offer the reader the possibility to gain a deeper understanding of the physical phenomena studied in this book. Appendices 1 and 2 relate, respectively, to the description of *quantum wires*, *quantum wells* and *quantum dots* of semiconductor materials. This description facilitates the connection with *potential wells* and *potential dots* studied in quantum mechanics. Moreover, these appendices make it possible to introduce the notions of *2D, 1D and 0D confinement*. Finally, Appendix 3 focuses on the detailed proof of the expression of the *transparency* of a potential barrier of height V_0 for a particle of energy $E > V_0$. This facilitates the introduction of the *resonance phenomenon*. A list of references and an index can be found at the end of the book.

I wish to thank Chrono Environement Laboratory at the Université Franche Comté de Besançon for their hospitality during my stay from September 1 to November 2, 2018 as a Visiting Professor. Many pages of this book were written during this period, which proved very favorable to this endeavor, both in terms of logistics and documentation. I would like to make a special mention to Jean-Emmanuel Groetz, Senior Lecturer at Chrono Environnement Laboratory, who was in charge of my Visiting Professor request file. I wish to express my warmest thanks to Elie Belorizky, Professor of Physics at Université Joseph Fourier de Grenoble (France), for his critical remarks and suggestions, which had a great contribution to improving the scientific quality of this work. Many corrections brought to this book have been made via telephone exchanges during my stay at the Université Franche

Comté de Besançon. I am expressing here my deep appreciation for him gracefully bearing the inherent expenses for the telephone calls related to this book review. Finally, I wish to address my deepest gratitude to Louis Marchildon, Professor of Physics (Emeritus) at the Université de Quebec à Trois Rivières (Canada), who spared no effort to review the entire book, and whose comments have enhanced the scientific quality of this work, whose foreword bears his signature. We started our collaboration in 2013, when he invited me to host a conference at the Hydrogen Research Institute (HRI). I am deeply grateful for his kind and very fruitful collaboration.

All human endeavor being subject to improvement, I remain open to and interested in critical remarks and suggestions that my readers can send me at the below-mentioned email.

Ibrahima SAKHO
aminafatima_sakho@yahoo.fr
October 2019

1

Schrödinger's Equation and its Applications

General objective

The general objective is to apply the Schrödinger equation to the study of simple physical systems.

Specific objectives

On completing this chapter, the reader should be able to:

– know the properties of the square-summable wave functions;

– know the boundary conditions imposed to any square-summable wave function;

– distinguish between a physical state in classical mechanics and in quantum mechanics;

– describe a physical quantity in quantum mechanics;

– define an operator;

– define an observable;

– give examples of operators and observables;

– know the correspondence principle or rule;

– define the Hamiltonian of a physical system;

– express the time-dependent Schrödinger equation;

– express the stationary Schrödinger equation;

For color versions of the figures in this book, see www.iste.co.uk/sakho/quantum2.zip.

– know the properties of the Schrödinger equation;

– integrate the Schrödinger equation for a free particle;

– integrate the Schrödinger equation for the ground state of the hydrogen atom;

– apply the Schrödinger equation to the study of quantum wells;

– apply the Schrödinger equation to the study of quantum dots;

– apply the Schrödinger equation to the study of potential barriers;

– apply the Schrödinger equation to the study of potential steps;

– define the probability current;

– define the reflection and transmission factors;

– define the reflection and transmission probabilities;

– provide an interpretation of the tunnel effect;

– describe the scanning tunneling microscope.

Prerequisites

– De Broglie plane wave.

– Heisenberg's uncertainty relations.

– Properties of trigonometric functions.

– Euler formulae.

– Integer series.

1.1. Physical state and physical quantity

1.1.1. *Dynamic state of a particle*

According to classical mechanics, the dynamic state of a particle is fully determined at each moment if the position $\vec{r}(x,y,z)$ and velocity or linear momentum $\vec{p}(p_x,p_y,p_z)$ of this particle are known. In particular, if its position and velocity at an instant $t = 0$ are known, it is possible to calculate, by solving the fundamental equation of dynamics, its dynamic state at a subsequent moment t and hence its trajectory.

Given the uncertainty principle, the notion of trajectory loses its meaning and a different approach must be adopted for the characterization of the dynamic state. The mathematical entities that can describe the dynamic states of the particle must reflect its wave-like nature. Hence, an orbital dynamic state of the particle is described by a generally complex wave function $\Psi(\vec{r},t)$.

1.1.2. *Physical quantities associated with a particle*

In classical mechanics, the measurable physical quantities associated with a particle such as kinetic or potential energy and angular momentum are expressed as functions of position variables x, y, z and linear momentum variables p_x, p_y, p_z. For example:

– its kinetic energy is written as $E_c = (p_x^2 + p_y^2 + p_z^2)/2m$;

– its orbital angular momentum with respect to a point O of the space is written as $\vec{\sigma} = \overrightarrow{OM} \wedge \vec{p}$.

In quantum mechanics, the measurable physical quantities are represented by Hermitian operators, as described in section 1.3.2. For example, for a given particle:

– operator $\vec{P} = \frac{\hbar}{i}\vec{\nabla}$ represents its linear momentum;

– operator $T = -\frac{\hbar^2}{2m}\vec{\nabla}^2$ represents its kinetic energy;

– operator $\vec{R}$ represents its position.

In contrast to classical mechanics, which does not distinguish between state and physical quantity, there is an essential difference between the two notions in quantum mechanics: a state is represented by a state vector, while a physical quantity is represented by an operator, which is generally denoted by A.

1.2. Square-summable wave function

1.2.1. *Definition, superposition principle*

As already explained above, the wave function describing the physical state of a particle is a complex function $\Psi(\vec{r},t)$ satisfying the normalization condition [4.49]. The set of square-summable wave functions constitutes the Hilbert space denoted by L^2 [COH 77, MAR 00, NEU 18].

If $\Psi_1(\vec{r},t)$ and $\Psi_2(\vec{r},t)$ are two square-summable wave functions and if λ_1 and λ_2 are two complex numbers, then any linear combination of these two functions is also a square-summable wave function:

$$\Psi(\vec{r},t) = \lambda_1 \Psi_1(\vec{r},t) + \lambda_2 \Psi_2(\vec{r},t) \qquad [1.1]$$

Relation [1.1] satisfies the *superposition principle*.

1.2.2. *Properties*

Generally speaking, for bound states there are discontinuous square-summable wave functions. Nevertheless, in quantum mechanics, the square-summable wave functions used have the following properties:

– they are continuous and indefinitely differentiable;

– their derivatives with respect to space variables are *continuous*, even at possible points of discontinuity of potentials;

– they are *zero at infinity* according to the normalization condition [4.49];

– they satisfy the *scalar product* of two functions defined in the Hilbert space.

Let $\Phi(\vec{r})$ and $\Psi(\vec{r})$ be two square-summable wave functions. By definition, the scalar product of $\Phi(\vec{r})$ and $\Psi(\vec{r})$ is the complex number denoted by (Ψ, Φ) and given by the relation:

$$(\Psi,\Phi) = \int_{-\infty}^{+\infty} \Psi^* \Phi \, d^3 r \qquad [1.2]$$

The scalar product uses the complex conjugate Ψ^* of the wave function Ψ.

If λ_1 and λ_2 are two complex numbers, the scalar product [1.2] has the properties:

$$\begin{cases} (\Phi,\Psi)^* = (\Psi,\Phi) \\ (\Phi,\lambda_1\Psi_1 + \lambda_2\Psi_2) = \lambda_1(\Phi,\Psi_1) + \lambda_2(\Phi,\Psi_2) \\ (\lambda_1\Phi_1 + \lambda_2\Phi_2,\Psi) = \lambda_1^*(\Phi_1,\Psi) + \lambda_2^*(\Phi_2,\Psi) \end{cases} \qquad [1.3]$$

According to properties [1.3], the scalar product is linear with respect to the second function of the pair and anti-linear with respect to the first function of the pair. The definition of the scalar product makes it possible to define the *norm of a square-summable wave function*. For $\Psi \equiv \Phi$, relation [1.2] becomes:

$$(\Psi,\Psi) = \int_{-\infty}^{+\infty} \Psi^* \Psi \, d^3 r = \int_{-\infty}^{+\infty} |\Psi|^2 \, d^3 r \qquad [1.4]$$

By definition, the norm of a wave function denoted by $\|\Psi\|$ is given by the following relation:

$$\|\Psi\| = \sqrt{(\Psi,\Psi)} = \sqrt{\int_{-\infty}^{+\infty} |\Psi|^2 \, d^3r} \geq 0 \qquad [1.5]$$

Equality [1.5] is satisfied when the wave function is zero.

1.3. Operator

1.3.1. *Definition of an operator, examples*

By definition, an operator denoted by A is a mathematical being whose action on a wave function Ψ transforms it into another wave function Φ. The transformation equation is written as follows:

$$A\Psi = \Phi \qquad [1.6]$$

Some operator examples are listed below:

– multiplication by x denoted by X: $X\Psi(x) = x\Psi(x) = \Phi(x)$;

– differentiation with respect to x denoted by $\partial/\partial x$;

$$\frac{\partial \Psi(x)}{\partial x} = \Psi'(x)$$

– parity denoted by Π:

$\Pi\Psi(x) = \Psi(x)$: if $\Psi(x)$ is even

or $\Pi\Psi(x) = -\Psi(x)$: if $\Psi(x)$ is odd.

1.3.2. *Hermitian operator*

Considering the scalar product of ψ and $A\Psi$, we have:

$$(\Psi A, \psi) = \int A^{\dagger}\Psi^*(\vec{r})\psi(\vec{r})\, d^3r \qquad [1.7]$$

Operator $A^{\dagger}$ (A dagger) is by definition the *adjoint operator* of A.

Moreover, an operator that is its own adjoint is called a *Hermitian operator* or a *self-adjoint operator*. Any Hermitian operator A verifies the relation $A = A^{\dagger}$. Given the properties [1.3] of the scalar product, any Hermitian operator verifies the property:

$$(\Psi, A\psi) = \int \Psi^*(\vec{r}) A\psi(\vec{r})\, d^3r = \left(\int \psi^*(\vec{r}) A\Psi(\vec{r})\, d^3r \right)^* \qquad [1.8]$$

The simple definition of a Hermitian operator will be explored in Chapter 3, after the introduction of Dirac notations and the notion of matrix element.

NOTE (HERMITIC OPERATOR AND HERMITIAN OPERATOR).– There are quantum mechanics works that feature the adjective *Hermitic*. The appropriate adjective is, nevertheless, *Hermitian*, for at least two reasons. First, as teaching experience indicates, students often confuse the words *hermitic* and *hermetic* (which the students are very familiar with). Second, many operators have been named after famous scientists who contributed to the development of quantum mechanics formalism. It is the case of Lagrangian, Laplacian, Hamiltonian, etc. The respective names of these operators honor the French naturalized Italian mathematician, mechanics scientist and astronomer **Joseph Louis comte de Lagrange** (1736–1813), the French mathematician, physicist, astronomer and politician **Pierre-Simon de Laplace** (1749–1827) and the Irish mathematician, physicist and astronomer **Sir William Rowan Hamilton** (1805–1865). To avoid the confusion with the quasi-homonymous adjective hermetic, it is wiser to use the adjective Hermitian, as a reference to the French mathematician **Charles Hermite** (1822–1901) (Box 1.1).

APPLICATION 1.1.–

Let A be a self-adjoint operator. Is the operator $B = iA$ Hermitian?

Solution. Let us find the adjoint operator of B: $B^\dagger = (iA)^\dagger = (i)^* A^\dagger = -iA \Rightarrow B^\dagger = -B$: operator B is not Hermitian.

Charles Hermite was a French mathematician. His work focused on the theory of numbers, quadratic forms, orthogonal polynomials, elliptic functions and differential equations. In quantum mechanics, Hermitian operators as well as Hermite polynomials, used in the study of the quantum harmonic oscillator, are mathematical concepts known as Hermitian in his honor.

In 1925, he developed in parallel to Schrödinger (see Box 1.3) the first theorization of quantum mechanics within matrix formalism (while Schrödinger adopted a rather wave-like approach by solving the differential equations). In 1927, Heisenberg stated the indeterminacy principle rejecting the notion of trajectory of a microscopic particle. He was awarded the Nobel Prize for physics in 1933 for his works in quantum mechanics.

Box 1.1. *Hermite (1822–1901)*

1.3.3. *Linear observable operator*

By definition, a linear operator is a mathematical being that establishes a linear correspondence between any wave function Ψ and another wave function Ψ'. If A is a linear operator, then:

$$\begin{cases} A\Psi = \Psi' \\ A(\lambda_1\Psi_1 + \lambda_2\Psi_2) = \lambda_1 A\Psi_1 + \lambda_2 A\Psi_2 \end{cases} \quad [1.9]$$

The foundation of physics relies on observation and experimentation or measurement. In quantum mechanics, any measurable physical quantity is associated with an operator, which is an *observable*.

An observable is defined as a Hermitian operator whose eigen functions (or eigen vectors, see Chapter 3) form a *complete set*. A set is complete to the extent that every square-summable wave function is written in only one way, as a convergent series expansion on the basis of the eigen functions of this observable. The fundamental observables based on which all the others are expressed in quantum mechanics are operators associated with the position $\vec{r}$, linear momentum $\vec{p}$ and the total mechanical energy E of a system (see section 1.3.4).

APPLICATION 1.2.–

Prove that the operator multiplication by z and the operator first derivative with respect to variable y are linear operators.

Solution.

– *Operator multiplication by z*: Using [1.9], we have:

$$\begin{cases} Z\Psi = z\Psi \\ Z(\lambda_1\Psi_1 + \lambda_2\Psi_2) = z(\lambda_1\Psi_1 + \lambda_2\Psi_2) \end{cases} \Rightarrow Z(\lambda_1\Psi_1 + \lambda_2\Psi_2) = \lambda_1 z\Psi_1 + \lambda_2 z\Psi_2$$

This gives:

$$Z(\lambda_1\Psi_1 + \lambda_2\Psi_2) = \lambda_1 Z\Psi_1 + \lambda_2 Z\Psi_2$$

– *Operator first derivative with respect to variable y*: Let d_y be the first derivative with respect to variable y. We have:

$$d_y(\lambda_1\Psi_1 + \lambda_2\Psi_2) = \frac{\partial}{\partial y}(\lambda_1\Psi_1 + \lambda_2\Psi_2)$$

This gives:

$$d_y(\lambda_1\Psi_1+\lambda_2\Psi_2)=\lambda_1\frac{\partial\Psi_1}{\partial y}+\lambda_2\frac{\partial\Psi_2}{\partial y}$$

Hence:

$$d_y(\lambda_1\Psi_1+\lambda_2\Psi_2)=\lambda_1 d_y\Psi_1+\lambda_2 d_y\Psi_2$$

1.3.4. *Correspondence principle, Hamiltonian*

In quantum mechanics, the principle according to which an observable A can be determined from classical mechanics quantities is governed by an empirical rule known as the *correspondence principle* [ATT 05] or *correspondence rule* [BAY 17]. All ambiguity should be removed before proceeding, given that the correspondence principle developed in this section differs from *Bohr's correspondence principle.*

Indeed, in 1923 Bohr formulated a heuristic principle known as *Bohr's correspondence principle*. This principle, which was very useful upon the start of quantum mechanics development, states that the *results of quantum mechanics must agree with those of classical mechanics at the limit of very large quantum numbers* (see exercise 3.7.7, Chapter 3, Volume 1). In other terms, when the discrete character of measurable quantities can be ignored, the results provided by quantum mechanics can be determined with very good approximation within the framework of classical mechanics. The applicability of this correspondence principle goes beyond quantum mechanics. This principle is also valid in relativistic mechanics. For example, when $v/c \ll 1$, Lorentz factor (equation [4.66], Chapter 4, Volume 1) $\gamma \approx 1$ and the laws of relativistic mechanics coincide with those of classical mechanics. This section takes a different approach to the formulation of the correspondence principle, since it employs the notion of observable, which was unknown during the development of Bohr's theory.

Before stating the correspondence principle, let us list the expressions of the observables associated with the physical quantities position $\vec{r}$, linear momentum $\vec{p}$ and energy E, which are the most commonly used in quantum mechanics. These are the following:

– position $\vec{r}$ $(x, y, z) \rightarrow$ position operator $\vec{R}$ (X, Y, Z);

– linear momentum $\vec{p}$ $(x, y, z) \rightarrow$ linear momentum operator $\vec{P}$ (P_x, P_y, P_z);

– potential energy $V(\vec{r}) \rightarrow$ potential energy operator $V(\vec{R})$;

– kinetic energy $E_c = p^2/2m$ → kinetic energy operator $T = \vec{P}^2/2m$;

– mechanical energy E → Hamiltonian H.

Let us note that the linear momentum operator and the Hamiltonian are, respectively, expressed as functions of the Laplacian and the operator first derivative with respect to time:

$$\vec{P} = i\frac{\hbar}{i}\vec{\nabla}\,;\ H = i\hbar\frac{\partial}{\partial t} \qquad [1.10]$$

In order to prove relations [1.10], let us consider a one-dimensional problem that analyzes the wave associated with a free particle that moves with a well-defined linear momentum $P = P_x$. In this case, De Broglie plane wave [4.1] can be written considering Planck–Einstein relations [2.54] as follows:

$$\Psi(x,t) = \Psi_0\, e^{i(px/\hbar - Et/\hbar)} \qquad [1.11]$$

Using expression [1.11], we determine the following partial derivatives (putting $\Psi(x, t) = \Psi$ in order to simplify):

$$\begin{cases} i\hbar\dfrac{\partial\Psi}{\partial t} = E\Psi \\ -\dfrac{\hbar^2}{2m}\dfrac{\partial^2\Psi}{\partial x^2} = \dfrac{P_x^2}{2m}\Psi = E\Psi \end{cases} \Rightarrow \begin{cases} H\Psi = E\Psi \\ P_x^2\Psi = \dfrac{\hbar^2}{i^2}\dfrac{\partial^2}{\partial x^2}\Psi \end{cases} \qquad [1.12]$$

This leads to:

$$H = i\hbar\frac{\partial}{\partial t}\mathbb{1};\ P_x = \frac{\hbar}{i}\frac{\partial}{\partial x} = \frac{\hbar}{i}\nabla_x \qquad [1.13]$$

Relations [1.10] are obtained if the expression of operator P_x is generalized to three dimensions.

In the relation [1.13], $\mathbb{1}$ designates the *identity operator* [COH 77, SAH 12]. This operator is also denoted by the symbol $\hat{I}$ [BAS 17]. The identity operator is often omitted and for simplicity purposes we can write:

$$H = i\hbar\frac{\partial}{\partial t} \qquad [1.14]$$

We can now formulate the correspondence principle so that it makes it possible to determine the expression of an observable from a classical expression:

> "The observable A ($\vec{R}$, $\vec{P}$, t) describing a physical quantity A ($\vec{r}$, $\vec{p}$, t) defined in classical mechanics is obtained by conveniently symmetrizing the classical expression and then by replacing $\vec{p}$ by $-i\hbar\nabla$ in the symmetrized expression".

Example: Let us determine the observable associated with the classical quantity $A(\vec{r},\vec{p}) = \vec{r}\cdot\vec{p}$.

It is worth noting that given the commutativity of the scalar product, we have:

$$A(\vec{r},\vec{p}) = \vec{r}\cdot\vec{p} = \vec{p}\cdot\vec{r} \quad [1.15]$$

On the other hand, $\vec{R}$ and $\vec{P}$ operators, which are associated with $\vec{r}$ and $\vec{p}$, respectively, are not always commutative. This follows from Heisenberg uncertainty principle. For example:

$XP_x \neq P_xX$ but $XP_y = P_yX$

Hence, in the general case, $\vec{R}\cdot\vec{P} \neq \vec{P}\cdot\vec{R}$.

From a classical point of view,

$$\vec{r}\cdot\vec{p} = \frac{1}{2}(\vec{r}\cdot\vec{p} + \vec{r}\cdot\vec{p}) \quad [1.16]$$

The symmetrization of the classical expression [1.16] leads to: 1/2 $(\vec{r}\cdot\vec{p} + \vec{p}\cdot\vec{r})$.

The observable $A(\vec{R},\vec{P})$ can therefore be written as:

$$A(\vec{R},\vec{P}) = \frac{1}{2}(\vec{R}\cdot\vec{P} + \vec{P}\cdot\vec{R}) = -\frac{i\hbar}{2}(\vec{R}\cdot\nabla + \nabla\cdot\vec{R}) \quad [1.17]$$

NOTE.– Commutation operator is a very important notion in quantum mechanics. This is why Chapter 3 is dedicated to its detailed study. We shall keep in mind for the time being that the scalar product of two operators is commutative provided that the physical quantities described by the two operators are simultaneously measurable.

APPLICATION 1.3.–

Find the expression of the observable describing the mechanical energy of a conservative system.

Solution. The mechanical energy of a conservative system is constant. It is given by the classical expression:

$$E = \frac{p^2}{2m} + V(r) \qquad [1.18]$$

The associated observable is the Hamiltonian H given by the quantum expression:

$$H = \frac{\vec{P}^2}{2m} + V(\vec{R}) = -\frac{\hbar^2}{2m}\Delta + V(\vec{R}) \qquad [1.19]$$

In the relations [1.19], **Δ** is the Laplacian, with $\nabla^2 = \Delta$.

Sir **William Rowan Hamilton** was an Irish mathematician, physicist and astronomer. He contributed to the development of optics, dynamics and algebra. He conducted significant researches for the development of analytical mechanics. The Hamiltonian operator or briefly the Hamiltonian involved in Schrödinger equation was named in his honor.

Box 1.2. *Hamilton (1805–1865)*

1.4. Evolution of physical systems

1.4.1. *Time-dependent Schrödinger equation*

In 1926, Schrödinger postulated the fundamental equation of quantum mechanics. According to this postulate, the evolution in time of a system is governed by the equation:

$$i\hbar\frac{\partial\Psi(\vec{r},t)}{\partial t} = H\Psi(\vec{r},t) \qquad [1.20]$$

In equation [1.20], H is the Hamiltonian observable associated with the total energy of the system. For time-dependent phenomena, the potential energy is a function of position and time. The Hamiltonian is written according to [1.19]:

$$H = -\frac{\hbar^2}{2m}\Delta + V(\vec{R}, t) \qquad [1.21]$$

Expression [1.21] shows that the Hamiltonian is a function of time. It is for this reason that equation [1.20] is known as *time-dependent Schrödinger equation*. Using [1.21], the partial differential equation [1.20] can be written in the following form:

$$i\hbar \frac{\partial \Psi(\vec{r}, t)}{\partial t} = \left(-\frac{\hbar^2}{2m}\Delta + V(\vec{r}, t) \right) \Psi(\vec{r}, t) \qquad [1.22]$$

1.4.2. *Stationary Schrödinger equation*

In physics, many systems are subjected to time-independent potentials. It is particularly the case of hydrogen-like systems, potential wells, potential barriers, quantum harmonic oscillator, etc.; the Schrödinger equation [1.20] has for these systems a particular form where $V(\vec{r}, t) = V(\vec{r})$. To establish this equation, the *variable separation method* will be used. For this purpose, particular solutions for equation [1.20] are sought for, writing the wave function as a product of a function of spatial coordinates $\Phi(\vec{r})$ and another time function $\chi(t)$:

$$\Psi(\vec{r}, t) = \Phi(\vec{r}) \times \chi(t) \qquad [1.23]$$

Using [1.23], the Schrödinger equation [1.22] can be written as follows:

$$i\hbar \Phi(\vec{r}) \frac{d\chi(t)}{dt} = \left[-\frac{\hbar^2}{2m}\nabla^2 + V(\vec{r}) \right] \Phi(\vec{r}) \times \chi(t) \qquad [1.24]$$

ATTENTION.– Simplification of both terms of equation [1.24] by $\Phi(\vec{r})$ should be avoided. Indeed, the right term contains the Laplacian of $\Phi(\vec{r})$, which is not equal to $\Phi(\vec{r})$. The division of both terms of [1.24] by $\Phi(\vec{r}) \times \chi(t)$ leads to:

$$i\hbar \frac{1}{\chi(t)} \frac{d\chi(t)}{dt} = \frac{1}{\Phi(\vec{r})} \left[-\frac{\hbar^2}{2m}\nabla^2 + V(\vec{r}) \right] \Phi(\vec{r}) \qquad [1.25]$$

Knowing that the term on the left side of [1.25] depends only on time and the one on the right side depends only on the variable $\vec{r}$, then the two terms are equal to a constant C. Moreover, each of these terms is equivalent to an energy. It can therefore be written as $C = E$. This leads to:

$$\begin{cases} i\hbar \dfrac{d\chi(t)}{dt} = E\chi(t) \\ \left[-\dfrac{\hbar^2}{2m}\nabla^2 + V(\vec{r}) \right] \Phi(\vec{r}) = E\Phi(\vec{r}) \end{cases} \qquad [1.26]$$

It can be noted that the term between square brackets in [1.26] contains the expression of the Hamiltonian [1.19] for conservative systems, which is:

$$H\Phi(\vec{r}) = E\Phi(\vec{r}) \qquad [1.27]$$

Equation [1.28] is known as the *stationary Schrödinger equation*. This equation makes it possible to solve many physical phenomena related to the behavior of time-independent potentials (see section 1.6). For this purpose, equation [1.27] is used in the form:

$$\left[-\frac{\hbar^2}{2m}\nabla^2 + V(\vec{r}) \right] \Phi(\vec{r}) = E\Phi(\vec{r}) \qquad [1.28]$$

Erwin Rudolf Josef Alexander Schrödinger was an Austrian physicist. In 1926, he postulated the non-relativistic wave equation describing the physical state of a system and opened the way for the development of the mathematical formalism of quantum mechanics. This wave equation, known as Schrödinger's equation in his honor, brought him the Nobel Prize for physics in 1933, which he shared with **Paul Dirac** (see Chapter 2, Box 2.1). In 1935, Schrödinger imagined the cat paradox, a thought experiment that was later called Schrödinger's cat, which is an evidence of the fracture between the *microscopic realm* (in which an object can simultaneously have several states) and the deterministic *macroscopic realm*.

Box 1.3. *Schrödinger (1887–1961)*

NOTE.– Schrödinger's cat is a thought experiment (a cat and a flask of poison) used by Schrödinger in his attempt to refute the Copenhagen interpretation (see note at the end of section 4.2.2, Chapter 4, Vol. 1) of quantum mechanics, which involved a simultaneously dead and alive cat. Shortly before submitting his thought experiment to publication, Schrödinger presented it to Einstein, who subsequently employed gunpowder and a nearby cat in the thought experiment. Schrödinger and Einstein thought that the possibility of a dead-alive cat proved that the Max Born interpretation of a wave function was incomplete. For more details on this experiment, the reader is invited to check the references [GRI 08, WIK 18].

1.4.3. *Evolution operator*

The first equation of the system [1.26] can be written in the following form:

$$\frac{d\chi(t)}{\chi(t)} = -i\frac{E}{\hbar}dt \tag{1.29}$$

This differential equation can be easily integrated and has the following solution:

$$\chi(t,t_0) = \chi(t_0)e^{-i\frac{E}{\hbar}(t-t_0)} \tag{1.30}$$

Let us consider χ (t_0) = 1 since this constant is not involved in the physical predictions that feature the density of probability. This gives:

$$\chi(t,t_0) = e^{-i\frac{E}{\hbar}(t-t_0)} \tag{1.31}$$

Solution [1.31] makes it possible to introduce an important operator denoted U for the determination of the wave function Ψ ($\vec{r}$, t) describing the evolution of a physical system based on the wave function Ψ ($\vec{r}$, t_0) at the initial moment t_0. Considering that $A = U$ and $\Phi = \chi$ in [1.6], we have:

$$U\Psi = \chi(t,t_0) = e^{-i\frac{E}{\hbar}(t-t_0)} \tag{1.32}$$

Knowing that the Hamiltonian H is the observable associated with the total energy E, the expression of operator U [1.31] can be deduced:

$$U(t,t_0) = e^{-i\frac{H}{\hbar}(t-t_0)} \tag{1.33}$$

By definition, operator U is known as *evolution operator* acting on the eigen function of H. The passage from Ψ ($\vec{r}$, t_0) to Ψ ($\vec{r}$, t) is expressed by the following relation:

$$\Psi(\vec{r}, t) = U(t, t_0)\,\Psi(\vec{r}, t_0) \tag{1.34}$$

Expression [1.33] is mentioned in Chapter 3 when studying conservative systems.

APPLICATION 1.4. –

By definition, A is a *unitary operator* if its adjoint coincides with its inverse. Prove that the evolution operator U is a unitary operator.

Solution. A being a unitary operator, then: $A^\dagger = A^{-1}$.

Using [1.33], we have:

$$U^\dagger U = UU^\dagger = 1 \Rightarrow U^\dagger = U^{-1} \tag{1.35}$$

1.5. Properties of Schrödinger's equation

1.5.1. *Determinism in the evolution of physical systems*

Schrödinger's equation [1.20] is a first-order partial differential equation with respect to time. The value of the wave function $\Psi(\vec{r}, t_0)$ at the initial instant t_0 being given, the wave function $\Psi(\vec{r}, t)$ at a given instant t can be found. There is no *indeterminism* in the evolution of physical systems in time. Indeterminism occurs during the measurement of a physical quantity on the considered system. During the measurement, the state vector denoted $|\Psi(t)\rangle$ undergoes an unpredictable jump due to what is known as *fundamental perturbation* [COH 77].

1.5.2. *Superposition principle*

Let $\Psi_1(\vec{r}, t)$ and $\Psi_2(\vec{r}, t)$ be two wave functions that are solutions of the Schrödinger equation [1.20]. Let us consider that at instant t_0, the state of the system is described by the wave function $\Psi(\vec{r}, t_0)$ such that:

$$\Psi(\vec{r}, t_0) = \lambda_1\Psi_1(\vec{r}, t_0) + \lambda_2\Psi_2(\vec{r}, t_0) \tag{1.36}$$

It should be reminded that in relation [1.36], λ_1 and λ_2 are complex numbers.

Then at a given instant t, the wave function describing the system is written as:

$$\Psi(\vec{r}, t) = \lambda_1\Psi_1(\vec{r}, t) + \lambda_2\Psi_2(\vec{r}, t) \tag{1.37}$$

Result [1.37] shows that any *linear combination* of wave functions that are solutions of the Schrödinger equation is also a solution of the same equation. Therefore Schrödinger's equation [1.20] satisfies the *superposition principle.*

For a *conservative system*, the Hamiltonian H is time independent. The passage from $\Psi(\vec{r}, t_0)$ to $\Psi(\vec{r}, t)$ is linear and is made by the evolution operator according to relation [1.34].

1.5.3. *Probability current density*

For a stationary wave function, the normalization condition [4.51] reflects the fact that the probability of finding the system at point r in space is equal to the unity. In other words, probability is conserved. This *probability conservation* involves the fact that the density of probability [4.49] is constant, even if the system evolves in time.

Let us consider a general case for the study of the principle of probability conservation. For this purpose, let us first recall the principle of *conservation of the electric charge.*

Let us consider a system of charged particles of *volume charge density* $\rho(\vec{r},t)$. Charge variation $dq = \rho(\vec{r}, t)\, dV$ in time corresponds to the flow of an electric current I through a cross-section dS limiting the volume dV so that $dq = Idt$. The global charge $Q = \int dq$ is nevertheless conserved. This principle of global conservation of the electric charge relies on a local conservation of charge reflected by the *continuity equation* [SAK 18]:

$$\frac{\partial\rho(\vec{r},t)}{\partial t} + \vec{\nabla}\cdot\vec{J}(\vec{r},t) = 0 \qquad [1.38]$$

In this relation, $\vec{J}(\vec{r},t)$ is the *current density* flux going out of dS, the surface being perpendicular to the current density vector.

In quantum mechanics, a *probability current density* vector $\vec{J}(\vec{r},t)$ is introduced, which satisfies an equation of type [1.38] reflecting a principle of *local probability conservation.* To illustrate this point, it is convenient to imagine a "probability flow". If the probability of finding the particle in the volume element dV around point $\vec{r}$ varies, it is because the flux of the probability current through surface dS limiting the volume dV is not zero. To establish the continuity equation satisfied by

$\vec{J}(\vec{r},t)$, the starting point is the time-dependent Schrödinger equation [1.22], which is reminded below:

$$i\hbar\frac{\partial\Psi(\vec{r},t)}{\partial t}=\left[-\frac{\hbar^2}{2m}\nabla^2+V(\vec{r})\right]\Psi(\vec{r},t) \qquad [1.22b]$$

The complex conjugate of this equation is:

$$-i\hbar\frac{\partial\Psi^*(\vec{r},t)}{\partial t}=\left[-\frac{\hbar^2}{2m}\nabla^2+V(\vec{r})\right]\Psi^*(\vec{r},t) \qquad [1.39]$$

Multiplying both sides [1.22] by Ψ* and [1.39] by –Ψ* [and putting Ψ= Ψ($\vec{r}$, t) for the sake of simplification], we have:

$$i\hbar\Psi^*\frac{\partial\Psi}{\partial t}=-\frac{\hbar^2}{2m}\Psi^*\nabla^2\Psi+V(\vec{r})\Psi^*\Psi \qquad [1.40]$$

$$i\hbar\Psi\frac{\partial\Psi^*}{\partial t}=\frac{\hbar^2}{2m}\Psi\nabla^2\Psi^*-V(\vec{r})\Psi\Psi^* \qquad [1.41]$$

The sum of these two equations is:

$$i\hbar\left(\Psi^*\frac{\partial\Psi}{\partial t}+\Psi\frac{\partial\Psi^*}{\partial t}\right)=-\frac{\hbar^2}{2m}\left(\Psi^*\nabla^2\Psi-\Psi\nabla^2\Psi^*\right) \qquad [1.42]$$

Arranging the member on the left side of equation [1.42], we get:

$$i\hbar\frac{\partial}{\partial t}(\Psi^*\Psi)=-\frac{\hbar^2}{2m}\left(\Psi^*\nabla^2\Psi-\Psi\nabla^2\Psi^*\right) \qquad [1.43]$$

Since the probability density satisfies the relation Ψ*Ψ = $|\Psi|^2$, then we have $\rho(\vec{r},t)=\Psi^*\Psi$. Equation [1.43] can then be written in the following form:

$$\frac{\partial\rho(\vec{r},t)}{\partial t}+\frac{\hbar}{2mi}\left(\Psi^*\nabla^2\Psi-\Psi\nabla^2\Psi^*\right)=0 \qquad [1.44]$$

Equation [1.44] is identical to [1.42] if we consider:

$$\vec{\nabla}\cdot\vec{J}(\vec{r},t)=\frac{\hbar}{2mi}\left(\Psi^*\nabla^2\Psi-\Psi\nabla^2\Psi^*\right) \qquad [1.45]$$

In order to deduce the expression of $\vec{J}(\vec{r},t)$ from relation [1.45], let us add to the term between brackets of the left-side member, the quantity $(\vec{\nabla}\Psi^*).(\vec{\nabla}\Psi)-(\vec{\nabla}\Psi^*).(\vec{\nabla}\Psi)$. We then obtain:

$$\vec{\nabla}\cdot\vec{J}(\vec{r},t)=\frac{\hbar}{2mi}\left[\Psi^*\nabla^2\Psi-\Psi\nabla^2\Psi^*+(\vec{\nabla}\Psi^*)\cdot(\vec{\nabla}\Psi)-(\vec{\nabla}\Psi^*)\cdot(\vec{\nabla}\Psi)\right] \quad [1.46]$$

This then gives:

$$\vec{\nabla}\cdot\vec{J}(\vec{r},t)=\frac{\hbar}{2mi}\left[(\vec{\nabla}\Psi^*).(\vec{\nabla}\Psi)+\Psi^*\nabla^2\Psi-(\vec{\nabla}\Psi).(\vec{\nabla}\Psi^*)-\Psi\nabla^2\Psi^*\right]$$

which is:

$$\vec{\nabla}\cdot\vec{J}(\vec{r},t)=\frac{\hbar}{2mi}\vec{\nabla}\cdot\left[\Psi^*\vec{\nabla}\Psi-\Psi\vec{\nabla}\Psi^*\right] \quad [1.47]$$

Therefore, the probability current density is written as:

$$\vec{J}(\vec{r},t)=\frac{\hbar}{2mi}\left[\Psi^*\vec{\nabla}\Psi-\Psi\vec{\nabla}\Psi^*\right] \quad [1.48]$$

Equation [1.45] can then be written as:

$$\frac{\partial\rho(\vec{r},t)}{\partial t}+\vec{\nabla}\cdot\vec{J}(\vec{r},t)=0 \quad [1.49]$$

The continuity equation [1.49] reflects the probability conservation.

The probability current density [1.49] is often expressed as a function of the three-dimensional linear momentum operator [ATT 08, BAY 17].

In one dimension q, probability current density [1.48] can be written as a function of the linear momentum operator $P_q=-i\hbar\nabla_q$ in the following form:

$$J_q=\frac{1}{2m}(\Psi^*P_q\Psi-\Psi P_q\Psi^*)=\frac{1}{m}\mathrm{Re}(\Psi^*P_q\Psi) \quad [1.50]$$

In relation [1.50], Re designates the *real part* of the complex number $(\Psi^*P_q\Psi)$. In three dimensions, we have:

$$\vec{J}(\vec{r},t) = \frac{1}{m}\mathrm{Re}\left[\Psi^*(\vec{r},t)\vec{P}\Psi(\vec{r},t)\right] \qquad [1.51]$$

In [1.51], the linear momentum operator is given by the first relation [1.10].

1.6. Applications of Schrödinger's equation

1.6.1. *Infinitely deep potential well*

The behavior of a particle confined in an infinitely deep potential well of width a [COH 77, GRI 95, PHI 03, MAR 00, STÖ 07, BEL 03, ATT 05, SAK 12, BAY 17] is studied. The profile of the potential energy $V(x)$ is shown in Figure 1.1.

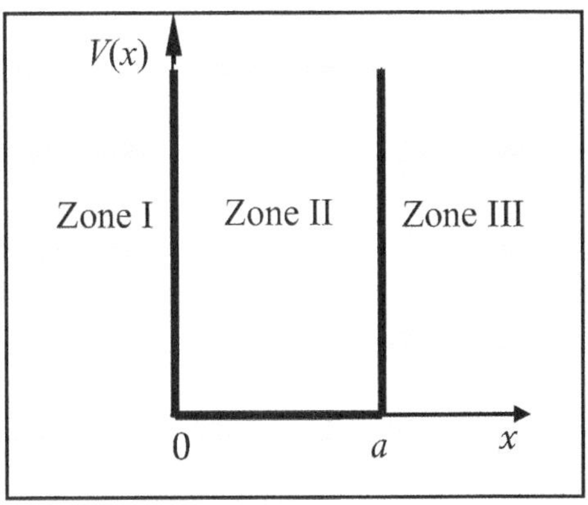

Figure 1.1. *Infinitely deep potential well of width a*

The potential energy function satisfies the following conditions:

$$V(x) = \begin{cases} 0, & 0 \le x \le a \\ +\infty, & x<0 \text{ and } x>a \end{cases} \qquad [1.52]$$

1.6.1.1. *Behavior of the particle*

The energy E of the particle is equal to:

$$E = \frac{p^2}{2m} + V(x)$$

This then leads to:

$$\frac{p^2}{2m} = E - V(x) \tag{1.53}$$

The above relation [1.53] shows that, if the potential is infinite, the kinetic energy of the particle becomes negative. In other terms, the speed of the particle becomes imaginary. This has no meaning in classical mechanics: the zone is impenetrable. From the quantum point of view, it can be proved that the wave function is zero in a space domain where the potential is infinite. Therefore, in the zones where the potential is infinite, the wave function satisfies the *boundary conditions*:

$$\Phi\,(0) = 0; \Phi\,(a) = 0 \tag{1.54}$$

1.6.1.2. *Analysis of Schrödinger's equation*

In the well, the potential energy is $V(x) = 0$. According to [1.54], we have:

$$E = \frac{p^2}{2m} > 0 \tag{1.55}$$

From a classical point of view, the particle is then executing oscillations between the limits $x = 0$ and $x = a$, its kinetic energy being $E_c = E$. From a quantum point of view, the state of the particle is governed by the stationary Schrödinger equation [1.28]. In one dimension, this is:

$$-\frac{\hbar^2}{2m}\frac{d^2\Phi(x)}{dx^2} = E\Phi(x) \tag{1.56}$$

The integration of the differential equation [1.56] makes it possible to determine the nature of the spectrum of the particle confined in the well. For this purpose, let us find the solutions to equation [1.56], which can be written as:

$$\frac{d^2\Phi(x)}{dx^2} + \frac{2mE}{\hbar^2}\Phi(x) = 0 \tag{1.57}$$

Given the Planck–Einstein relations [4.3] $p = \hbar k$, [1.57] can be written as:

$$E = \frac{\hbar^2 k^2}{2m} \Rightarrow k^2 = \frac{2mE}{\hbar^2} \tag{1.58}$$

Using [1.58], equation [1.57] can be written in the form:

$$\frac{d^2\Phi(x)}{dx^2}+k^2\,\Phi(x)=0 \qquad [1.59]$$

One solution of equation [1.59] is of the type:

$$\Phi(x)=A\sin kx+B\sin kx \qquad [1.60]$$

1.6.1.3. *Energy quantization*

The boundary conditions [1.54] require the wave function $\Phi\,(x)$ to be continuous at the well connection points (in $x = 0$ and in $x = a$). Therefore:

– continuity in $x = 0 \Rightarrow \Phi_{\mathrm{I}}\,(0) = \Phi_{\mathrm{II}}\,(0) = 0 \Rightarrow B = 0$;

– continuity in $x = a \Rightarrow \Phi_{\mathrm{II}}\,(a) = \Phi_{\mathrm{III}}\,(a) = 0 \Rightarrow A \sin ka = 0$. Hence:

$$ka = n\pi \Leftrightarrow k_n = \frac{n\pi}{a} \qquad [1.61]$$

Result [1.61] reflects the quantization of the wave vector norm. Consequently, the energy of the particle is also quantized according to [1.58]. Hence:

$$E_n = \frac{\hbar^2\pi^2}{2ma^2}n^2 \qquad [1.62]$$

Therefore, the spectrum of the particle is discrete. In [1.62], n is the quantum number, which is strictly positive, since the energy is not zero, given the uncertainty principle. Indeed, if $n = 0$, $E = 0$. The linear momentum p is therefore zero. Given Heisenberg's first uncertainty relation [4.59], the position of the particle is infinite, which is impossible, since it is confined in the well.

Let us rewrite expression [1.62] in the form:

$$E_n = n^2 E_1\ ;\ E_1 = \frac{\hbar^2\pi^2}{2ma^2} \qquad [1.63]$$

Figure 1.2 represents the discrete spectrum of the particle for several energy levels. The values of energy E_n are proportional to the ground state energy E_1.

Result [1.63] reveals the essential difference between the physical predictions of classical mechanics and those of quantum mechanics.

From a classical point of view, the energy E of the particle is continuous (from 0 to infinity since the speed of the particle is under no restriction).

From the perspective of quantum predictions, due to the physical properties of the wave function, the energy E of the particle can only take discrete values in the well. Moreover, for the one-dimensional *quantum harmonic oscillator*, the energy is given by relation [3.263]:

$$E_n = \hbar\omega\left(n + \frac{1}{2}\right) \Rightarrow \Delta E_n = \left|E_{n\pm 1} - E_n\right| = \hbar\omega \qquad [1.64]$$

The energy gap between these two consecutive levels of the quantum harmonic oscillator is therefore constant (the energy levels are equidistant). On the other hand, the energy gap between two consecutive levels of the particle confined in the well is equal to $(2n + 1)$, as shown in Figure 1.2. This is due to the fact that the energy [1.63] of the particle varies with the square of the quantum number n.

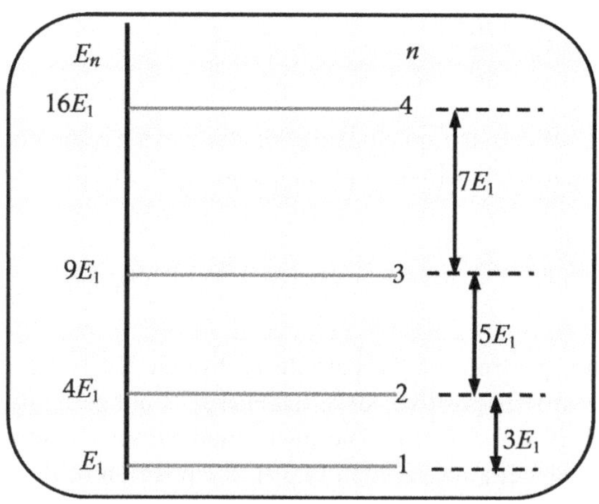

Figure 1.2. *Discrete spectrum of a particle confined to an infinitely deep potential well*

1.6.1.4. *Expression of the normed wave function*

Using [1.61] and considering that B is zero, [1.60] gives:

$$\Phi_n(x) = A\sin\left(\frac{n\pi}{a}x\right) \qquad [1.65]$$

Given the normalization condition of the wave function:

$$\int_0^a \|A\|^2 \sin^2\left(\frac{n\pi}{a}x\right)dx = 1$$

Considering the transformation $\cos 2q = 1 - 2\sin^2 q$, $(q = n\pi x/a)$, the integration of the previous equation gives:

$$\|A\|^2 \frac{a}{2} = 1 \Rightarrow A = \sqrt{\frac{2}{a}} \qquad [1.66]$$

In summary, the normed wave function satisfies the following equations:

$$\Phi_n(x) = \begin{cases} 0 & x < 0 \\ \sqrt{\frac{2}{a}} \sin\left(\frac{n\pi x}{a}\right), & 0 < x < a \\ 0 & x > 0 \end{cases} \qquad [1.67]$$

1.6.1.5. *Expression of the probability density*

In zone II, the density of the probability of particle presence is given by the square of the probability amplitude $\Phi_{II}(x)$. Using [1.67], we have:

$$\rho_n(x) = \frac{2}{a}\sin^2\left(\frac{n\pi}{a}x\right) \qquad [1.68]$$

Expression [1.68] shows that the probability density $\rho_n(x)$ is zero at the well connection points ($x = 0$ and $x = a$). Consequently, it has a maximum between 0 and a. The maximum of this probability density is obtained for:

$$\sin^2\left(\frac{n\pi}{l}x\right) = 1$$

For $x > 0$, we have:

$$\frac{n\pi}{a}x = (2k+1)\frac{\pi}{2} \Rightarrow x_n = \frac{a}{2n(2k+1)} \qquad [1.69]$$

Let us consider the particular case when the integer $k = 0$. For the ground state ($n = 1$) and the first excited state ($n = 2$), the maxima of probability density $\rho_n(x)$ correspond to $x_1 = l/2$ and $x_2 = l/4$.

The plots of the wave function $\Phi_n(x)$ and of the probability density $\rho_n(x)$ are shown in Figure 1.3 below for the ground level $n = 1$ and for the first two excited levels $n = 2$ and 3.

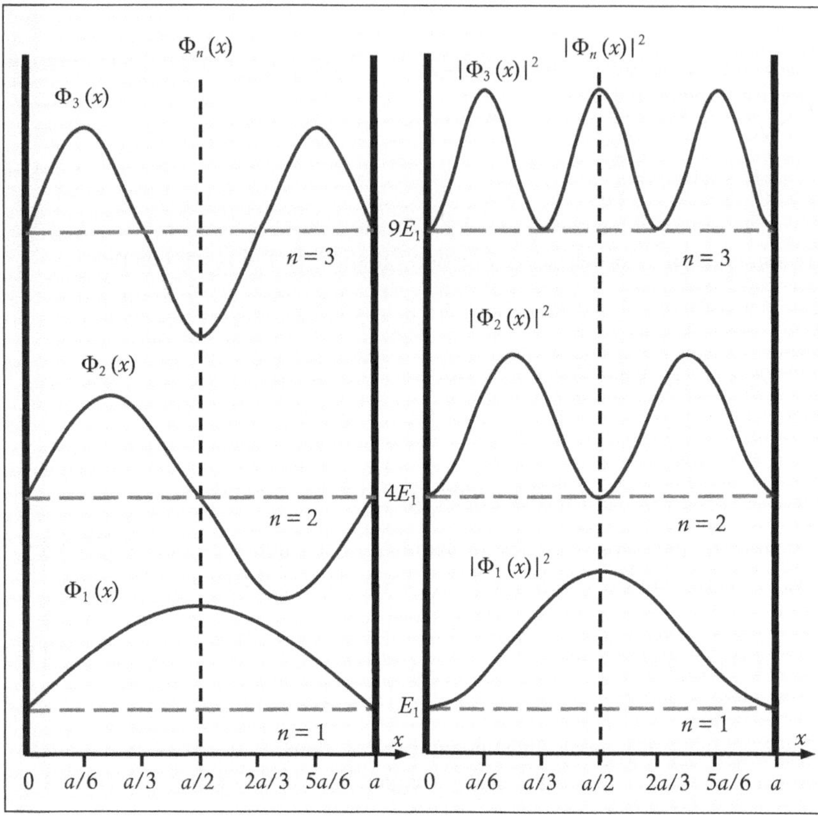

Figure 1.3. *Variations of the wave function and of the probability density of a particle confined in an infinitely deep potential well*

1.6.2. *Potential step*

This section focuses on the study of a particle coming from $-\infty$ and heading to a *potential step* [COH 77; PÉR 86; SIV 86; GRI 95; PHI 03; ATT 05; STÖ 07; SAK 12; BAY 17] of height V_0 (Figure 1.4). This is a rectangular potential barrier (Figure 1.5) of width $a \to \infty$. The kinetic energy of the particle is denoted by E.

Here, we study the behavior of the particle through the potential step when the energy $E > V_0$, then when $E < V_0$.

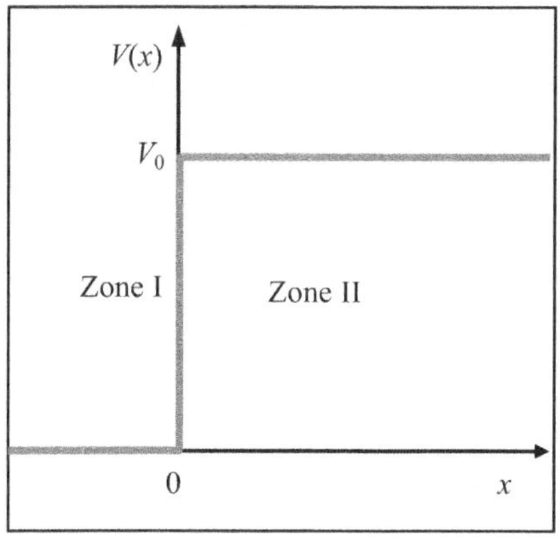

Figure 1.4. *Potential step of height V_0*

1.6.2.1. *Case of $E > V_0$*

1.6.2.1.1. Classical and quantum predictions

From a classical point of view, the particle flies over the step and carries on along its path: its motion is rectilinear with a speed drop at the transition point of abscissa $x = 0$. From the quantum mechanics point of view, the state of the particle is described by a wave function. There is a non-zero probability for the particle to be reflected or transmitted through the potential step. To analyze these purely quantum phenomena, let us consider Schrödinger equation [1.28] in one dimension:

$$\left[-\frac{\hbar^2}{2m}\frac{d^2}{dx^2}+V(x)\right]\Phi(x)=E\Phi(x) \qquad [1.70]$$

In zone I, the potential is zero. Hence, according to [1.70]:

$$\frac{\hbar^2}{2m}\frac{d^2\Phi_{\text{I}}(x)}{dx^2}+E\Phi_{\text{I}}(x)=0 \qquad [1.71]$$

In zone II, the potential is equal to V_0. Equation [1.70] then yields:

$$\frac{\hbar^2}{2m}\frac{d^2\Phi_{\text{II}}(x)}{dx^2}+(E-V_0)\Phi_{\text{II}}(x)=0 \qquad [1.72]$$

1.6.2.1.2. General solutions

The general solutions of equations [1.71] and [1.72] respectively, are as follows:

Zone I

$$\Phi_{\mathrm{I}}(x) = A\exp(ik_{\mathrm{I}}x) + B\exp(-ik_{\mathrm{I}}x) \quad [1.73]$$

with:

$$k_{\mathrm{I}} = \sqrt{\frac{2mE}{\hbar^2}} \quad [1.74]$$

Zone II

$$\Phi_{\mathrm{II}}(x) = C\exp(ik_{\mathrm{II}}x) + D\exp(-ik_{\mathrm{II}}x) \quad [1.75]$$

with:

$$k_{\mathrm{II}} = \sqrt{\frac{2m(E - V_0)}{\hbar^2}} \quad [1.76]$$

1.6.2.1.3. Amplitude reflection and transmission factors

Expression [1.73] of the wave function is a superposition of an *incident plane wave* exp $(ik_{\mathrm{I}}x)$ and a *reflected plane wave* exp $(-ik_{\mathrm{I}}x)$. This proves that the particle can either be reflected or transmitted through the potential barrier; there is no similarity in classical mechanics. The objective is to find the expression of the reflection and transmission probabilities, and then to verify the law of probability conservation.

In zone II, there is no *backward wave*. Therefore, the coefficient $D = 0$. To summarize, only the following solutions should be considered:

$$\begin{cases} \Phi_{\mathrm{I}}(x) = A\exp(ik_{\mathrm{I}}x) + B\exp(-ik_{\mathrm{I}}x) \\ \\ \Phi_{\mathrm{II}}(x) = C\exp(ik_{\mathrm{II}}x) \end{cases} \quad [1.77]$$

Let us use the boundary conditions imposed to the wave function $\Phi'_{\mathrm{i}}(x)$ and to its first derivative $\Phi'_{\mathrm{i}}(x) = d\Phi\,(x)/dx$ in $x = 0$:

$$\begin{aligned} &\Phi_{\mathrm{I}}(0) = \Phi_{\mathrm{II}}(0) \\ &\Phi'_{\mathrm{I}}(0) = \Phi'_{\mathrm{II}}(0) \end{aligned} \quad [1.78]$$

Using [1.77], the boundary conditions [1.78] lead to the following system:

$$\begin{cases} A + B = C \\ k_{\mathrm{I}}(A - B) = k_{\mathrm{II}}C \end{cases} \quad [1.79]$$

Arranging [1.79], we express B and C as functions of A. Therefore, we obtain:

$$B = \frac{k_{\mathrm{I}} - k_{\mathrm{II}}}{k_{\mathrm{I}} + k_{\mathrm{II}}} A \quad [1.80]$$

$$C = \frac{2k_{\mathrm{I}}}{k_{\mathrm{I}} + k_{\mathrm{II}}} A \quad [1.81]$$

By definition, the *amplitude reflection factor* denoted r and the *amplitude transmission factor* denoted d of the waves at the level of the barrier result from the following relations:

$$r = \frac{B}{A}; \; d = \frac{C}{A} \quad [1.82]$$

Using relations [1.80] and [2.82], we obtain:

$$r = \frac{k_{\mathrm{I}} - k_{\mathrm{II}}}{k_{\mathrm{I}} + k_{\mathrm{II}}}; \; d = \frac{2k_{\mathrm{I}}}{k_{\mathrm{I}} + k_{\mathrm{II}}} \quad [1.83]$$

1.6.2.1.4. Reflection probability R and transmission probability T

By definition, the *reflection probability* R and the *transmission probability* T of the particle are given by the following relations:

$$R = \frac{|B|^2}{|A|^2} = |r|^2; \; T = \frac{k_{\mathrm{II}}}{k_{\mathrm{I}}} \times \frac{|C|^2}{|A|^2} = \frac{k_{\mathrm{II}}}{k_{\mathrm{I}}} \times |d|^2 \quad [1.84]$$

Using expressions [1.80] and [1.81], we finally obtain:

$$R = \left(\frac{k_{\mathrm{I}} - k_{\mathrm{II}}}{k_{\mathrm{I}} + k_{\mathrm{II}}} \right)^2; \; T = \frac{4k_{\mathrm{I}}k_{\mathrm{II}}}{(k_{\mathrm{I}} + k_{\mathrm{II}})^2} \quad [1.85]$$

The transmission probability T [SAK 12, BAY 17] is also often known as *barrier permeability* or *barrier transparency* [SIV 86, BEL 03], or as the barrier transmission coefficient [COH 77, STÖ 07].

1.6.2.1.5. Probability conservation

Let us find the sum of reflection and transmission probabilities using [1.85]. We have:

$$R+T=\frac{(k_{\rm I}-k_{\rm II})^2}{(k_{\rm I}+k_{\rm II})^2}+\frac{4k_{\rm I}k_{\rm II}}{(k_{\rm I}+k_{\rm II})^2}$$

The expansion of the first term of the right-hand member leads to:

$$R+T=\frac{k_{\rm I}^2+k_{\rm II}^2-2k_{\rm I}k_{\rm II}+4k_{\rm I}k_{\rm II}}{(k_{\rm I}+k_{\rm II})^2}=\frac{k_{\rm I}^2+k_{\rm II}^2+2k_{\rm I}k_{\rm II}}{(k_{\rm I}+k_{\rm II})^2}=1$$

Hence:

$$R+T=1 \qquad [1.86]$$

Therefore, from the quantum perspective, the particle is either reflected or transmitted, while from the classical mechanics perspective it moves past the barrier, so no reflection takes place. Among others, result [1.86] reflects the *law of conservation of mass*.

1.6.2.2. *Case when $E < V_0$*

1.6.2.2.1. Value of the reflection factor, evanescent wave

When $E < V_0$, the quantity $k_{\rm II}$ becomes imaginary according to [1.76]. Similarly to geometric optics, *total reflection* takes place. Consequently, the probability is $R = 1$. Indeed, if we consider $k_{\rm II} = i\rho$, the amplitude reflection factor [1.83] can be written as:

$$r=\frac{k_{\rm I}-i\rho}{k_{\rm I}+i\rho}\Rightarrow T=|r|^2=\frac{|k_{\rm I}-i\rho|^2}{|k_{\rm I}+i\rho|^2}=1$$

Though total reflection occurs, the wave transmitted in zone II is not zero: it is transformed into a wave known as an *evanescent wave of low depth of penetration*. To establish the expression of this wave, we consider $k_{\rm II} = i\rho$. Using [1.77], we obtain:

$$\Phi_{\rm II}(x)=C\exp(-\rho x) \qquad [1.87]$$

with:

$$\rho=\sqrt{\frac{2m(V_0-E)}{\hbar^2}} \qquad [1.88]$$

1.6.2.2.2. Expression of the depth of penetration

The depth of penetration of the evanescent wave is the distance l_p at which the density of probability decreases by $1/e$ [SIV 86, SAK 12]. According to [1.87], the density of probability is:

$$D(x) = C^2 \exp(-2\rho x) \quad [1.89]$$

For $x = l_p$, $\exp(-2\rho l_p) = 1/e = \exp(-1)$. Therefore, $2\rho l_p = 1$, or $l_p = 1/2\rho$. Using [1.88], we finally find (knowing that $\hbar = h/2\pi$):

$$l_p = \frac{h}{4\pi\sqrt{2m(V_0 - E)}} \quad [1.90]$$

1.6.2.3. *Conclusion*

Result [1.90] shows that the wave penetrates zone II even though it undergoes total reflection.

It is worth noting the existence of a factor k_{II}/k_I in the expression [1.84] of the probability of transmission T. The origin of this factor should be clarified. For this purpose, let us express the *probability current densities* for the process of reflection and transmission of the particle using [1.73] and the expressions [1.75] of the wave functions $\Phi_I(x)$ and $\Phi_{II}(x)$. The complex conjugates of these functions can be written as follows:

$$\begin{cases} \Phi^*_I(x) = A\exp(-ik_I x) + B\exp(ik_I x) \\ \Phi^*_{II}(x) = C\exp(-ik_{II} x) \end{cases} \quad [1.91]$$

These relations are used in the calculation of the probability current in zones I and II. We obtain respectively:

$$J_I = \frac{\hbar}{2mi}\left[\Phi_I{}^*\frac{d}{dx}\Phi_I - \Phi_I\frac{d}{dx}\Phi_I{}^*\right] \quad [1.92]$$

$$J_{II} = \frac{\hbar}{2mi}\left[\Phi_{II}{}^*\frac{d}{dx}\Phi_{II} - \Phi_{II}\frac{d}{dx}\Phi_{II}{}^*\right] \quad [1.93]$$

Let us calculate the products between brackets involved in relations [1.92] and [1.93]. We obtain:

$$\begin{cases} \dfrac{d\Phi_I}{dx} = iAk_I\exp(ik_I x) - ik_I B\exp(-ik_I x) \\ \dfrac{d\Phi_{II}}{dx} = ik_{II} C\exp(ik_{II} x) \end{cases} \quad [1.94]$$

Similarly:

$$\begin{cases} \dfrac{d\Phi^*_{\mathrm{I}}}{dx} = -iAk_{\mathrm{I}} \exp(-ik_{\mathrm{I}}x) + ik_{\mathrm{I}}B\exp(ik_{\mathrm{I}}x) \\ \dfrac{d\Phi^*_{\mathrm{II}}}{dx} = -ik_{\mathrm{II}}C\exp(-ik_{\mathrm{II}}x) \end{cases} \qquad [1.95]$$

Taking [1.91] and [1.94] into account, this leads to:

$$\begin{cases} \Phi^*_{\mathrm{I}}\dfrac{d\Phi_{\mathrm{I}}}{dx} = i|A|^2 k_{\mathrm{I}} - ik_{\mathrm{I}}AB\exp(-2ik_{\mathrm{I}}x) + ik_{\mathrm{I}}AB\exp(2ik_{\mathrm{I}}x) - i|B|^2 k_{\mathrm{I}} \\ \\ \Phi_{\mathrm{I}}\dfrac{d\Phi^*_{I}}{dx} = -i|A|^2 k_{\mathrm{I}} + ik_{\mathrm{I}}AB\exp(2ik_{\mathrm{I}}x) - ik_{\mathrm{I}}AB\exp(-2ik_{\mathrm{I}}x) + i|B|^2 k_{\mathrm{I}} \end{cases}$$

$$\begin{cases} \Phi^*_{\mathrm{II}}\dfrac{d\Phi_{\mathrm{II}}}{dx} = ik_{\mathrm{II}}|C|^2 \\ \Phi_{\mathrm{II}}\dfrac{d\Phi^*_{\mathrm{II}}}{dx} = -ik_{\mathrm{II}}|C|^2 \end{cases}$$

Using these relations, the probability currents in zones I and II are written according to the probability currents [1.92] and [1.93]:

$$J_{\mathrm{I}} = \frac{\hbar k_{\mathrm{I}}}{m}\left[|A|^2 - |B|^2\right]; J_{\mathrm{II}} = \frac{\hbar k_{\mathrm{II}}}{m}|C|^2 \qquad [1.96]$$

The expression of J_{I} is the sum of these two terms. The first term corresponds to the incident probability current J_{Ii} (due to the incident wave) and the second term corresponds to the reflected probability current J_{Ir} (due to the reflected wave). Hence:

$$J_{\mathrm{I}i} = \frac{\hbar k_{\mathrm{I}}}{m}|A|^2 ; J_{\mathrm{I}r} = \frac{\hbar k_{\mathrm{I}}}{m}|B|^2 \qquad [1.97]$$

By definition, the *reflection probability* R is equal to the ratio of the *reflected probability current* J_{Ir} to the *incident probability current* J_{Ii}. If we use the results [1.97], we have:

$$R = \frac{J_{\mathrm{I}r}}{J_{\mathrm{I}i}} \Rightarrow R = \frac{|B|^2}{|A|^2} \qquad [1.98]$$

This is the first of relations [1.84]. Similarly, the *transmission probability* T is defined as the ratio of the *transmitted probability current* J_{IIr} to the incident probability current J_{Ii}. Hence:

$$T = \frac{J_{\mathrm{II}r}}{J_{\mathrm{I}i}} \Rightarrow T = \frac{k_{\mathrm{II}}}{k_{\mathrm{I}}} \times \frac{|C|^2}{|A|^2} \qquad [1.99]$$

This is the second of relations [1.84].

In the particular case of $k_{\mathrm{I}} = k_{\mathrm{II}} = k$, $T = 1$ and $R = 0$ according to [1.85]. A further consequence is that the coefficient $B = 0$. Therefore, the wave functions [1.77] are identical, since $A = C$ according to [1.99]:

$$\begin{cases} \Phi_{\mathrm{I}}(x) = A\exp(ikx) \\ \\ \Phi_{\mathrm{II}}(x) = C\exp(ikx) \end{cases} \Rightarrow \Phi_{\mathrm{I}}(x) = \Phi_{\mathrm{II}}(x) = A\exp(ikx) \qquad [1.100]$$

This is in agreement with the predictions of classical mechanics: the particle moves past the barrier without being reflected. Figure 1.5 shows the variations of the density of probability of presence $\rho\,(x) = |\Phi\,(x)|^2$ in the two cases considered ($E < V_0$ and $E > V_0$).

From a classical perspective, the particle is reflected for $E < V_0$, while from a quantum perspective the density of probability of presence $\rho\,(x)$ is not zero in the zone II forbidden by classical mechanics.

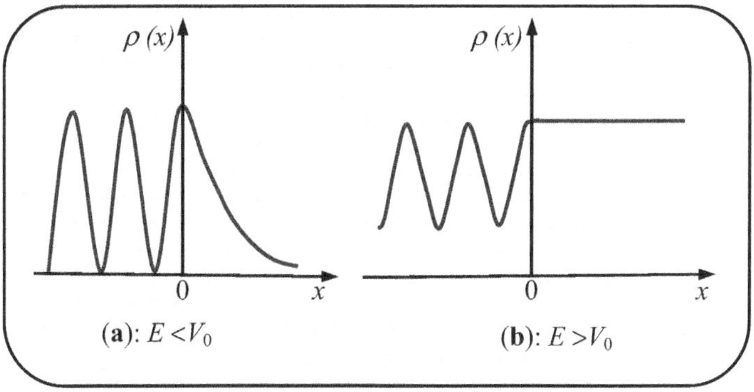

Figure 1.5. *Variations of the density of probability of presence $\rho\,(x)$ of a particle of energy E encountering a potential step of height V_0 (a) for $E < V_0$ and (b) for $E > V_0$*

Nevertheless, the density of probability decreases exponentially with x and becomes negligible when x is above the penetration length l_p [1.90] of the evanescent wave (Figure 1.5(a)): the particle is then definitely reflected. When the energy $E > V_0$, the particle is transmitted according to the classical predictions.

From a quantum perspective, the density of probability of presence $\rho(x)$ is constant in zone II and the particle has a non-zero probability R to go back according to [1.86]. Nevertheless, for $E >> V_0$ so that the height of the barrier can be ignored, then $k_I \approx k_{II}$ and the transmission probability $T \approx 1$ according to [1.85]: the particle is then transmitted according to the classical predictions, as shown in Figure 1.6 indicating the variation of probabilities T and R for $E < V_0$ and for $E > V_0$.

– At low energy ($E << V_0$), the coefficient $k_I \to 0$ according to relation [1.74]. Consequently $R \to 1$ and $T \to 0$ according to [1.85]: reflection is then total;

– for $E = V_0$, $k_{II} = 0$: $R = 1$ and $T = 0$. Reflection is then total;

– at high energy ($E >> V_0$), $k_I \to k_{II}$; $R \to 0$ and $T \to 1$: the particle is then transmitted according to the classical predictions.

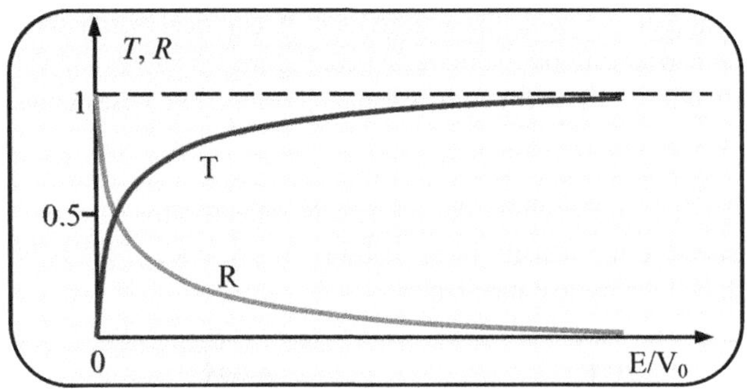

Figure 1.6. *Variations of the transmission probability T and of the reflection probability R depending on E/V₀*

1.6.3. *Potential barrier, tunnel effect*

Let us consider a *rectangular potential barrier* [COH 77, PÉR 86, SIV 86, GI 95, MAR 00, PHI 03, BEL 03, ATT 05, STÖ 07, SAK 12, BAY 17] of height V_0 and width a. A particle whose total energy is $E < V_0$ moves toward the barrier from a point of abscissa $x < 0$. The profile of the barrier is schematically represented in Figure 1.7. The potential described is such that:

$$V(x)=\begin{cases}0, & x<0\\ V_0, & 0\le x\le a\\ 0, & x>a\end{cases}$$

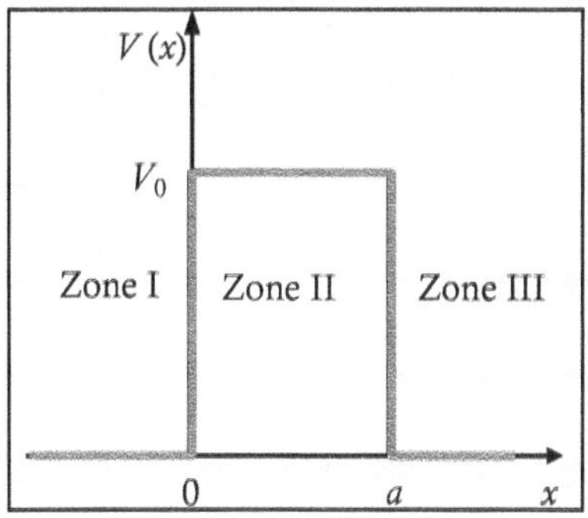

Figure 1.7. *Rectangular potential barrier of height V_0 and width a*

Our objective is to describe the behavior of the particle in contact with the barrier and to analyze the Schrödinger equation in the three zones: I, II and III. Then the case of $0 < E < V_0$ is considered. The case of $E > V_0$ will be dealt with in Appendix 3.

1.6.3.1. *Classical and quantum descriptions, Schrödinger's equation*

From the classical perspective, the particle is reflected: it hits the barrier and turns back with the same initial speed. From a quantum perspective, the particle behaves entirely differently. To describe the physical phenomena involved, we analyze the Schrödinger equation in the three zones: I, II and III. We obtain, respectively:

$$\frac{d^2\Phi_{\text{I}}(x)}{dx^2}+\frac{2mE}{\hbar^2}\Phi_{\text{I}}(x)=0 \qquad [1.101]$$

$$\frac{d^2\Phi_{\text{II}}(x)}{dx^2}+\frac{2m}{\hbar^2}(E-V_0)\Phi_{\text{II}}(x)=0 \qquad [1.102]$$

$$\frac{d^2\Phi_{\text{III}}(x)}{dx^2}+\frac{2mE}{\hbar^2}\Phi_{\text{III}}(x)=0 \qquad [1.103]$$

We consider:

$$k^2 = \frac{2mE}{\hbar^2};\ k'^2 = -\frac{2m(V_0 - E)}{\hbar};K^2 = \frac{2m(V_0 - E)}{\hbar} \qquad [1.104]$$

Let us write the solutions $\Phi_I(x)$, $\Phi_{II}(x)$ and $\Phi_{III}(x)$ of the above equations.

Zone I

The solution to equation [1.101] has the form:

$$\Phi_I(x) = Ae^{ikx} + Be^{-ikx} \qquad [1.105]$$

Zone II

Equation [1.102] has the following solution:

$$\Phi_{II}(x) = Ce^{ik'x} + De^{-ik'x} \qquad [1.106]$$

Since $E<V_0$, $k' = iK$ according to [1.104]. Solution [1.106] is then written as:

$$\Phi_{II}(x) = Ce^{-Kx} + De^{Kx} \qquad [1.107]$$

Zone III

The solution to equation [1.103] has the following form:

$$\Phi_{III}(x) = Fe^{ikx} + Ge^{-ikx} \qquad [1.108]$$

Since there is no reflected wave in zone III, then $G = 0$. Hence:

$$\Phi_{III}(x) = Fe^{ikx} \qquad [1.109]$$

1.6.3.2. *Expression of the barrier transparency*

Let us summarize the previous solutions to Schrödinger's equation:

$$\begin{cases} \Phi_I(x) = Ae^{ikx} + Be^{-ikx} \\ \Phi_{II}(x) = Ce^{-Kx} + De^{Kx} \\ \Phi_{III}(x) = Fe^{ikx} \end{cases} \qquad [1.110]$$

We now express the boundary conditions in $x = 0$ and then in $x = a$. We obtain:

$$\Phi_{I}(0) = \Phi_{II}(0);\ \Phi_{II}(a) = \Phi_{III}(a)$$

$$\Phi'_{I}(0) = \Phi'_{II}(0);\ \Phi'_{II}(a) = \Phi'_{III}(a) \qquad [1.111]$$

$$\Phi_i'(x) = \frac{d\Phi_i(x)}{dx}$$

Using [1.110] and [1.111], we have:

$$x=0:\begin{cases} A+B=C+D \\ ikA-ikB=-KC+KD \end{cases} \qquad [1.112]$$

$$x=a:\begin{cases} Ce^{-Ka}+De^{Ka}=Fe^{ika} \\ -KCe^{-Ka}+KDe^{Ka}=ikFe^{ika} \end{cases} \qquad [1.113]$$

The *probability of particle transmission* from zone I to zone III is given by the ratio: $T = |F|^2/|A|^2$. The proof of the expression of T is quite lengthy. It will be given in Appendix 3, as the focus here is on its physical interpretation. We then obtain:

$$T=\frac{4E(V_0-E)}{4E(V_0-E)+V_0^2 sh^2\left[\frac{\sqrt{2m(V_0-E)}}{\hbar}a\right]} \qquad [1.114]$$

1.6.3.3. *Tunnel effect*

Result [1.114] expresses the probability of the particle crossing the barrier. This partial transmission is known as the *tunnel effect*. Adopting a view according to which the particle goes through an already existing tunnel (for example, tunnels dug underground and used by subways or trains) should be avoided. This is a purely quantum effect reflecting the fact that a part of the incident wave is transmitted in the barrier and another part is reflected.

To clarify this, let us prove that the transparency T of the barrier decreases exponentially with its width a. For this purpose, we express T as a function of Ka using [1.104] and [1.114]. We have:

$$T=\frac{4E(V_0-E)}{4E(V_0-E)+V_0^2 sh^2[Ka]} \qquad [1.115]$$

Given the approximation $Ka >> 1$, the hyperbolic sine is sh$Ka \approx \exp(Ka)/2$. Equation [1.115] is then written as:

$$T \approx \frac{4E(V_0 - E)}{4E(V_0 - E) + \frac{V_0^2 e^{2Ka}}{4}} \qquad [1.116]$$

Since $Ka >> 1$, the first term of the denominator of [1.116] is negligible compared to the second term. Hence:

$$T \approx \frac{16E(V_0 - E)}{V_0^2} e^{-2Ka} \approx T_0 e^{-2Ka} \qquad [1.117]$$

with:

$$T_0 = \frac{16E(V_0 - E)}{V_0^2} \qquad [1.118]$$

The approximated result [1.117] shows that the tunnel effect decreases exponentially with the width a of the barrier and also with the mass m of the particle, since according to [1.105], we have:

$$K = \frac{\sqrt{2m(V_0 - E)}}{\hbar} = \alpha \times \sqrt{m} \qquad [1.119]$$

with:

$$\alpha = \frac{\sqrt{2(V_0 - E)}}{\hbar}$$

Relation [1.119] makes it possible to clarify the sensitivity of a particle to the tunnel effect.

As shown by expression [1.118], the tunnel effect decreases exponentially with the mass m of the particle via the constant K. The larger the mass of the particle, the faster the decrease of T to zero. Consequently, the intensity of the tunnel effect grows with the decrease in particle mass. Therefore, for the same value $(V_0 - E)$, an electron of mass m has a higher probability to generate a tunnel than the (smaller) proton of mass m_p $(m_p/m \approx 1{,}836)$.

1.6.3.3.1. General conclusion

The tunnel effect cannot be perceived for macroscopic objects. Figure 1.8 shows the variation of the density of probability $\rho(x) = |\Phi(x)|^2$ in zones I, II and III. The density of probability of presence $\rho(x)$ decreases exponentially and becomes zero in zone III beyond $x > l_p$.

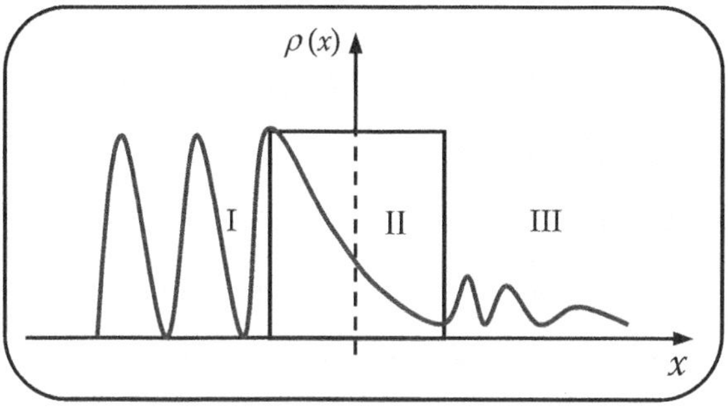

Figure 1.8. *Exponential decrease of the density of probability of presence ρ (x) of a particle that crosses a potential barrier by tunnel effect*

1.6.3.4. *Penetration length of a particle by tunnel effect*

Let us give an order of magnitude for a particle to cross a potential barrier by tunnel effect. We consider the case of the electron and proton. This makes it possible to theoretically confirm the assertion according to which the electron has a higher probability to cross the barrier by tunnel effect. The *depth of penetration* of the *evanescent wave* is given by [1.119], which is worth recalling:

$$l_p = \frac{h}{4\pi\sqrt{2m(V_0 - E)}} \qquad [1.119b]$$

Let us consider an electron and a proton of similar energy E = 1 eV each encountering a potential barrier of width a = 1 Å and height V_0 = 2 eV (the condition $E < V_0$ should be respected). Let us calculate the depth of penetration of the evanescent wave and the probability for the electron or proton to cross the barrier by tunnel effect. For the numerical applications: $h = 6.63 \times 10^{-34}$ J·s.

1.6.3.4.1. Length of penetration

– *for the electron* ($m = 9.1 \times 10^{-31}$ kg), we have: l_{pe} = 0.98 Å ≈ a = 1 Å;

– *for the proton* ($m_p = 1.67 \times 10^{-27}$ kg), we have: l_{pp} = 0.023 Å << a = 1 Å.

These results prove that the electron penetrates the barrier more than the proton; the length of penetration of the evanescent wave in the barrier is approximately 43 times ($l_{pe}/l_{pp} \approx 43$) larger for the electron than for the proton.

1.6.3.4.2. Transmission coefficient

Let us express the depth l_p as a function of K taking into account [1.119]. We have:

$$l_p = \frac{h}{4\pi\sqrt{2m(V_0 - E)}} = \frac{h}{4\pi\hbar K} = \frac{1}{2K} \Rightarrow K = 1/2l_p \quad [1.120]$$

Using [1.117] and [1.120], we have:

$$T \approx \frac{1}{1 + \frac{V_0^2}{16E(V_0 - E)} e^{a/l_p}} \quad [1.121]$$

Knowing that $E = 1$ eV and $V_0 = 2$ eV, formula [1.121] can be written as:

$$T \approx \frac{1}{1 + \frac{1}{4} e^{a/l_p}} \quad [1.122]$$

– *for the electron*: $l_{pe} = 0.98$ Å $\approx a = 1$Å $\Rightarrow T = 0.59 \approx 60\%$;

– *for the proton*: $l_{pe} = 0.023$ Å $\Rightarrow T \approx 5.2 \times 10^{-19} \approx 0\%$. [1.123]

We take $l_{pe}/l_{pp} \approx 43$. Hence, if $E = 1$ eV and $V_0 = 2$ eV, the results below show that while the electron has a 60% probability to cross the barrier by tunnel effect, the probability for the proton to cross the barrier by the same effect is extremely low.

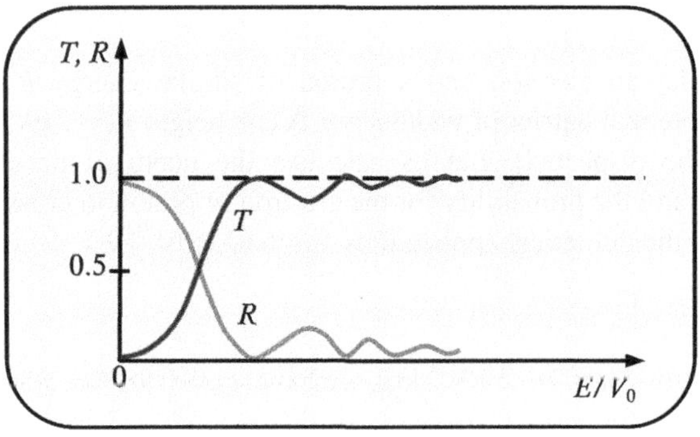

Figure 1.9. *Variations of the transmission probability T and the reflection probability R depending on E/V₀*

Figure 1.9 shows the variations of transmission probability T and reflection probability R depending on the ratio E/V_0. If $E >> V_0$ probability $T \approx 1$, the particle is certainly transmitted according to the predictions of classical mechanics.

1.6.4. *Quantum dot*

Let us consider a particle confined in a *quantum dot* [PÉR 86, STÖ 07, SAK 12] of edge a (Figure 1.10). This confinement makes it possible to define a cubic potential such that:

$$V(x) = \begin{cases} 0, & 0 < x < a; 0 < y < a; 0 < z < a; \\ \infty, & elsewhere \end{cases}$$

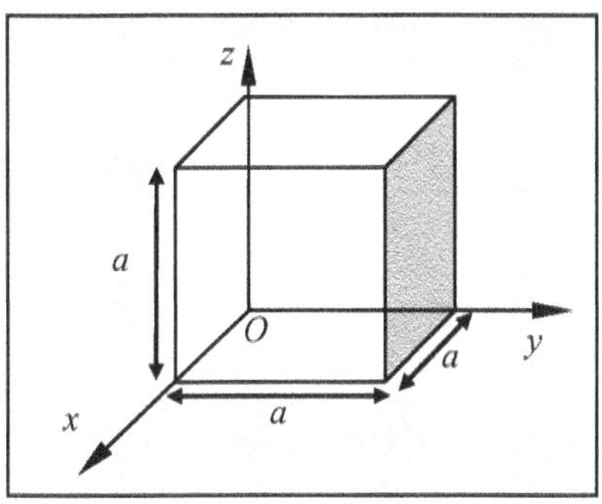

Figure 1.10. *Quantum dot of edge a*

Similar to the case of the two-dimensional rectangular potential, in the case of the cubic potential, the total energy $E = E_x + E_y + E_z =$ constant. Moreover, the global wave function $\Psi\,(x, y, z)$ describing the state of the particle in the dot is the product of three functions $\Phi(x)$, $\psi(y)$ and $\chi(z)$, which is:

$$\Psi\,(x, y, z) = \Phi(x) \times \psi(y) \times \chi(z) \qquad [1.124]$$

The transformation [1.124] is known as *ansatz*.

The objective here is to determine the spectrum of the particle in the dot and the explicit expression of the global wave function [1.124].

1.6.4.1. *The Schrödinger equation, normed wave function*

The density of the probability of finding the particle at a point $M(x, y, z)$ in the dot is given by the square of the amplitude of probability $\Psi(x, y, z)$. The potential is infinite outside the dot, $\Psi(x, y, z) = 0$. Consequently, the density of the probability of presence of the particle is zero outside the quantum dot. The Schrödinger equation in the dot is a three-dimensional equation. The potential being zero in the dot, this equation can be written according to [1.28]:

$$-\frac{\hbar^2}{2m}\left(\frac{\partial^2\Psi(x,y,z)}{\partial x}+\frac{\partial^2\Psi(x,y,z)}{\partial y}+\frac{\partial^2\Psi(x,y,z)}{\partial z}\right)=E\Psi(x,y,z) \qquad [1.125]$$

Using the *variable separation method* in [1.124], we obtain:

$$-\frac{\hbar^2}{2m}\left(\psi(y)\chi(z)\frac{d^2\Phi(x)}{dx}+\Phi(x)\chi(z)\frac{d^2\psi(y)}{dy}+\Phi(x)\psi(y)\frac{d^2\chi(z)}{dz}\right)=E\Phi(x)\psi(y)\chi(z) \qquad [1.126]$$

Dividing both terms of this equation by $\Phi(x) \times \psi(y) \times \chi(z)$, we have:

$$-\frac{\hbar^2}{2m}\left(\frac{1}{\Phi(x)}\frac{d^2\Phi(x)}{dx}+\frac{1}{\psi(y)}\frac{d^2\psi(y)}{dy}+\frac{1}{\chi(z)}\frac{d^2\chi(z)}{dz}\right)=E$$

Knowing that $E = E_x + E_y + E_z =$ constant, this equation can be written as:

$$-\frac{\hbar^2}{2m}\left(\frac{1}{\Phi(x)}\frac{d^2\Phi(x)}{dx}+\frac{1}{\psi(y)}\frac{d^2\psi(y)}{dy}+\frac{1}{\chi(z)}\frac{d^2\chi(z)}{dz}\right)=E_x+E_y+E_z \qquad [1.127]$$

By identification, we find:

$$\begin{cases}\dfrac{d^2\Phi(x)}{dx}+\dfrac{2mE_x}{\hbar^2}\Phi(x)=0\\[2ex] \dfrac{d^2\psi(y)}{dy}+\dfrac{2mE_y}{\hbar^2}\psi(y)=0\\[2ex] \dfrac{d^2\chi(z)}{dz}+\dfrac{2mE_y}{\hbar^2}\chi(z)=0\end{cases} \qquad [1.128]$$

The solutions to equations [1.128] are identical to solution [1.62] in the case of a particle confined in a potential well of width a. Hence:

$$\begin{cases} \Phi_n(x) = \sqrt{\frac{2}{a}} \sin\left(\frac{n\pi x}{a}\right) \\ \Psi_\mu(y) = \sqrt{\frac{2}{a}} \sin\left(\frac{\mu\pi y}{a}\right) \\ \chi_q(z) = \sqrt{\frac{2}{a}} \sin\left(\frac{q\pi z}{a}\right) \end{cases} \qquad [1.129]$$

In these expressions, n, μ and ν designate the quantum numbers characterizing the state of the particle along directions x, y and z, respectively. The normed wave function [1.124] $\Psi_{n\mu\nu}(x, y, z)$ can then be written as:

$$\Psi_{n\mu\nu}(x,y,z) = \sqrt{\frac{8}{a^3}} \sin\left(\frac{n\pi}{a}x\right) \sin\left(\frac{\mu\pi}{a}y\right) \sin\left(\frac{\nu\pi}{a}z\right) \qquad [1.130]$$

In [1.130], n, μ and ν have the minimal values $n = \mu = \nu = 1$. The value 0 should be excluded, otherwise the wave function is zero in the dot. Moreover, the value 0 (involving $E_0 = 0$) is forbidden by the uncertainty principle, as explained above in the case of the infinitely deep potential well.

1.6.4.2. *Spectrum of the particle, degeneracy of the energy levels*

Similar to the case of the rectangular potential, for each of the three degrees of freedom of the particle, the total energy is quantized and given by [1.62]. Therefore, for the three dimensions, we have:

$$E_n = \frac{\hbar^2\pi^2}{2ma^2} n^2 \; ; E_\mu = \frac{\hbar^2\pi^2}{2ma^2} \mu^2 \; ; E_\nu = \frac{\hbar^2\pi^2}{2ma^2} \nu^2 \qquad [1.131]$$

The total energy is then written as:

$$E_{n\mu\nu} = \frac{\hbar^2\pi^2}{2ma^2} \left(n^2 + \mu^2 + \nu^2\right) \qquad [1.132]$$

Expression [1.133] shows that the different quantum states for which the sum $n^2 + \mu^2 + \nu^2$ = constant have the same energy. There is therefore a degeneracy of the energy levels of the particle confined in the cubic box. It is worth recalling that the *degree* (*order*) *of degeneracy of a given energy level* $E_{n\mu q}$ is equal to the number of different quantum states that have the same energy. We illustrate this degeneracy in the case of the ground state and for specific combinations for which $n^2 + \mu^2 + \nu^2 = 6$ and 9. Table 1.1 summarizes the results obtained.

n	μ	ν	$E_{n\mu\nu}$	Degree of degeneracy (g)
1	1	1	$3E_0$	$g = 1$: non-degenerate ground level
1	1	2	$6E_0$	$g = 3$: three times degenerate excited level
1	2	1	$6E_0$	
2	1	1	$6E_0$	
1	2	2	$9E_0$	$g = 3$: three times degenerate excited level
2	1	2	$9E_0$	
2	2	1	$9E_0$	
2	2	2	$12E_0$	$g = 1$: non-degenerate excited level

Table 1.1. *Degeneracy of the energy levels of a particle confined in a cubic box. Only a few levels are presented*

1.6.5. *Ground state energy of hydrogen-like systems*

The quantization of the energy of the *hydrogen atom* has been established by Bohr since 1913. Our objective here is to find the expression of the ground state energy of hydrogen-like systems using Schrödinger's equation. $V(r)$ designates the potential energy of electron–nucleus interaction. Moreover, the stationary wave function describing the state of the electron is denoted by $\Psi(r,\theta,\varphi)$. Therefore, the stationary Schrödinger's equation should be integrated in *spherical coordinates*. In this system of coordinates, the Laplacian is given by the following expression:

$$\Delta = \frac{1}{r^2}\frac{\partial}{\partial r}\left(r^2\frac{\partial}{\partial r}\right) + \frac{1}{r^2 \sin\theta}\frac{\partial}{\partial\theta}\left(\frac{1}{\sin\theta}\frac{\partial}{\partial\theta}\right) + \frac{1}{r^2\sin^2\theta}\frac{\partial^2}{\partial^2\varphi} \quad [1.133]$$

1.6.5.1. *Schrödinger's equation*

Using Schrödinger's equation [1.28], we have:

$$\left[-\frac{\hbar^2}{2m}\Delta + V(\vec{r})\right]\Psi(\vec{r},\theta,\varphi) = E\Psi(\vec{r},\theta,\varphi) \quad [1.134]$$

For the stationary states, the wave function is the product of the radial wave function $\Phi(r)$ and the angular wave function $Y(\theta,\varphi)$ (in this notation the angular momentum quantum number and the orbital magnetic quantum number have been omitted; the spherical harmonics are denoted by $Y_\ell^m(\theta,\varphi)$). Hence:

$$\Psi(\vec{r},\theta,\varphi) = \Phi(r) \times \mathrm{Y}(\theta,\varphi) \qquad [1.135]$$

Taking into account this form of the wave function, equation [1.134] can be written as:

$$\left[-\frac{\hbar^2}{2m}\Delta + V(r)\right]\Phi(r) \times \mathrm{Y}(\theta,\varphi) = E\Phi(r) \times \mathrm{Y}(\theta,\varphi) \qquad [1.136]$$

Considering the ground state of a hydrogen-like system for which the 1s orbital has spherical symmetry, the global wave function [1.134] is independent of θ and φ and the function $\mathrm{Y}(\theta,\varphi)$ is constant. It only depends on the *radial coordinate r*. Consequently, the Laplacian [1.133] acts only on the radial part of the wave function. Hence, dividing [1.136] by $\mathrm{Y}(\theta,\varphi)$ in order to separate the angular and radial variables, we obtain:

$$\left[-\frac{\hbar^2}{2m}\Delta_r + V(r)\right]\Phi(r) = E\Phi(r) \qquad [1.137]$$

In this equation, the Laplacian is purely radial and is written according to [1.133]:

$$\Delta_r = \frac{1}{r^2}\frac{d}{dr}\left(r^2\frac{d}{dr}\right) \qquad [1.138]$$

Using [1.138], equation [1.137] becomes:

$$-\frac{\hbar^2}{2m}\frac{1}{r^2}\frac{d}{dr}\left(r^2\frac{d\Phi(r)}{dr}\right) + V(r)\Phi(r) = E\Phi(r)$$

After arrangement, we have:

$$\frac{1}{r^2}\frac{d}{dr}\left(r^2\frac{d\Phi(r)}{dr}\right) + \frac{2m}{\hbar^2}[E - V(r)]\Phi(r) = 0 \qquad [1.139]$$

Moreover, for a hydrogen-like system, the potential energy resulting from the Coulomb interaction between nucleus (+*Ze*) and electron (–*e*) is given by the relation:

$$V(r) = -k\frac{Ze^2}{r} \qquad [1.140]$$

Inserting [1.140] in equation [1.139], we have:

$$\frac{1}{r^2}\frac{d}{dr}\left(r^2\frac{d\Phi(r)}{dr}\right)+\frac{2m}{\hbar^2}\left(E+\frac{kZe^2}{r}\right)\Phi(r)=0$$

Hence:

$$\frac{d^2\Phi(r)}{dr^2}+\frac{2}{r}\frac{d\Phi(r)}{dr}+\frac{2m}{\hbar^2}\left(E+\frac{kZe^2}{r}\right)\Phi(r)=0$$

Or:

$$\frac{d^2\Phi(r)}{dr^2}+\frac{2}{r}\frac{d\Phi(r)}{dr}+\left(\frac{2mE}{\hbar^2}+\frac{2}{r}\frac{kZme^2}{\hbar^2}\right)\Phi(r)=0 \quad [1.141]$$

Let us consider:

$$\sigma=\frac{2mE}{\hbar^2};\rho=\frac{kZme^2}{\hbar^2} \quad [1.142]$$

The differential equation [1.141] can then be written in the form:

$$\frac{d^2\Phi(r)}{dr^2}+\frac{2}{r}\frac{d\Phi(r)}{dr}+\left(\sigma+\frac{2\rho}{r}\right)\Phi(r)=0 \quad [1.143]$$

1.6.5.2. *Ground state energy*

Knowing the wave function Φ (r) for the ground state makes it possible to determine the parameters ρ and σ and then deduce the expression of energy E according to [1.142]. One of the simplest solutions to equation [1.143] with a finite value for $r = 0$ and tending to zero for $r \to \infty$ has the following form [SIV 86, SAK 12]:

$$\Phi\ (r)=e^{-\mu r} \quad [1.144]$$

The first and second derivatives of this function can be written as:

$$\frac{d\Phi(r)}{dr}=-\mu e^{-\mu r};\frac{d^2\Phi(r)}{dr^2}=\mu^2 e^{-\mu r} \quad [1.145]$$

Inserting results [1.145] in [1.143], after arrangement we find:

$$(\mu^2 + \sigma) + \frac{1}{r}(2\rho - 2\mu) = 0 \qquad [1.146]$$

Equation [1.146] is verified for any value of variable r. Consequently, the terms between brackets in [1.146] are simultaneously zero. Hence:

$$\begin{cases} (\mu^2 + \sigma) = 0 \Rightarrow \sigma = -\mu^2 \\ (2\rho - 2\mu) = 0 \quad \Rightarrow \rho = \mu \end{cases} \qquad [1.147]$$

We deduce from these relations: $\sigma = -\rho^2$. According to [1.142], we then have:

$$\frac{2mE}{\hbar^2} = -\left(\frac{kZme^2}{\hbar^2}\right)^2$$

Or in the end:

$$E_1 = -\frac{k^2Z^2me^4}{2\hbar^2} \qquad [1.148]$$

Result [1.148] actually corresponds to the ground state energy of the hydrogen-like systems [3.24] if $n = 1$. If $Z = 1$, we find the expression of the ground state energy of the hydrogen atom known since 1913. This result is one of the strong confirmations of the validity of the Schrödinger equation postulated in 1926, which was 13 years after the elaboration of Bohr's theory on the hydrogen atom.

1.7. Exercises

1.7.1. *Exercise 1 – Probability current density*

The objective of this exercise is to express the *probability current density* in one dimension. For this purpose, we consider the *generalized coordinate* q (x, y or z) and the wave function $\Psi(q, t)$ of an arbitrary system (particle). Operator Q is associated with q.

(1) Express the Hamiltonian H of the system, then write the equations corresponding to the action of operators Q and $V(Q, t)$ on the wave function $\Psi(q, t)$.

(2) Let ρ be the probability density. Prove that:

$$\frac{\partial \rho}{\partial t} = \frac{\hbar}{2im}\left[\Psi\left(\frac{\partial^2 \Psi^*}{\partial q^2}\right) - \left(\frac{\partial^2 \Psi}{\partial q^2}\right)\Psi^*\right]$$

(3) Given the following relations:

$$\begin{cases} \dfrac{\partial}{\partial q}\left(\Psi \dfrac{\partial \Psi^*}{\partial q}\right) = \dfrac{\partial \Psi}{\partial q}\dfrac{\partial \Psi^*}{\partial q} + \Psi\left(\dfrac{\partial^2 \Psi^*}{\partial q^2}\right) \\ \dfrac{\partial}{\partial q}\left(\Psi^* \dfrac{\partial \Psi}{\partial q}\right) = \dfrac{\partial \Psi^*}{\partial q}\dfrac{\partial \Psi}{\partial q} + \Psi^*\left(\dfrac{\partial^2 \Psi}{\partial q^2}\right) \end{cases}$$

Prove the equation:

$$\frac{\partial \rho}{\partial t} = -\frac{\hbar}{2im}\frac{\partial}{\partial q}\left(\Psi^* \frac{\partial \Psi}{\partial q} - \frac{\partial \Psi^*}{\partial q}\Psi\right)$$

(4) Then prove that the equation for probability conservation can be written as:

$$\frac{\partial \rho(q,t)}{\partial t} + \frac{\partial J(q,t)}{\partial q} = 0$$

In this relation, $J(q, t)$ is a quantity to be defined.

1.7.2. *Exercise 2 – Heisenberg's spatial uncertainty relations*

This exercise focuses on the proof of Heisenberg's spatial uncertainty relations. For this purpose, let us consider the root mean square deviations Δx and Δp_x. The remaining two relations relative to y and z coordinates will be deduced by analogy. To simplify the study, we consider a one-dimensional problem and we choose the origin O of coordinates at the point of abscissa $\langle x \rangle = 0$ so that the mean of the linear momentum is $\langle p \rangle = 0$. Consequently:

$$\langle p \rangle = m\langle v \rangle = m\frac{d\langle x \rangle}{dt} = 0$$

The calculations will be done using the wave function $\Psi = \Psi(x)$ that is assumed normed to unity.

For all practical purposes, we have [CHP 78]:

$$\int \left| ux\Psi + w\frac{d\Psi}{dx} \right|^2 dx \geq 0 \,;\; \left(\Psi * \frac{d\Psi}{dx} \right) \Bigg|_{-\infty}^{+\infty} = 0$$

In the above inequality, *u* and *w* are auxiliary variables.

(1) Recall the definition of the average $\langle x \rangle$. Then deduce the root mean square deviations Δx and Δp_x by analogy.

(2) We consider:

$$A = \int x^2 \Psi\Psi * dx \,;\; B = -\int x \frac{d}{dx}(\Psi\Psi*) dx \,;\; C = \int \frac{d\Psi *}{dx} \frac{d\Psi}{dx} dx$$

Prove the inequality:

$$Au^2 - uwB + Cw^2 \geq 0.$$

Clarify the sign of A, with supporting rationale.

(3) Find the values of A and B. Then deduce the inequality verified by AC.

(4) Find the expression of C.

(5) Use the above results to deduce Heisenberg's uncertainty relations.

1.7.3. *Exercise 3 – Finite-depth potential step*

We consider a finite-depth potential step (Figure 1.11).

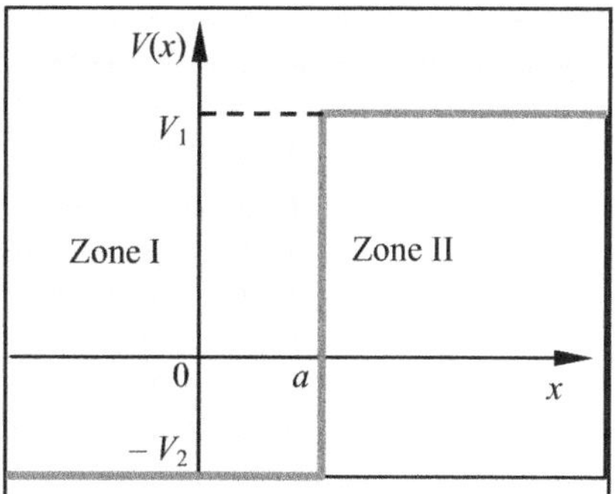

Figure 1.11. *Finite-height potential step*

A particle of energy E moves toward the potential step from a point of abscissa $x < 0$. The potential as described meets the following conditions:

$$V(x) = \begin{cases} V_1, & x > a \\ -V_2, & x < a \end{cases}$$

This exercise focuses on the study of the behavior of the particle in the following two cases:

$$E > V_1$$

and:

$$-V_2 < E < V_1$$

(1) *First case*: $E > V_1$.

(1.1) Prove that the states of the particle are stationary states. Write the Schrödinger equation in zones I ($x < a$) and II ($x > a$).

(1.2) Deduce the solutions $\Phi_I(x)$ and $\Phi_{II}(x)$ in zones I and II, respectively. We consider:

$$q^2 = \frac{2m(E+V_2)}{\hbar^2} \text{ for the solution in zone I}$$

$$\rho^2 = \frac{2m(E-V_1)}{\hbar^2} \text{ for the solution in zone II}$$

(1.3) Express the transmission probability T and the reflection probability R.

(1.4) Find $R + T$. Conclude by comparing the classical and quantum predictions relative to the behavior of the particle.

(2) *Second case*: $-V_2 < E < V_1$. Answer the same questions as for (1.1), (1.2), (1.3) and (1.4) above. As applicable, the expression of the evanescent wave and the emerging purely quantum effect shall be specified.

1.7.4. Exercise 4 – Multistep potential

Let us consider a multistep potential (Figure 1.12).

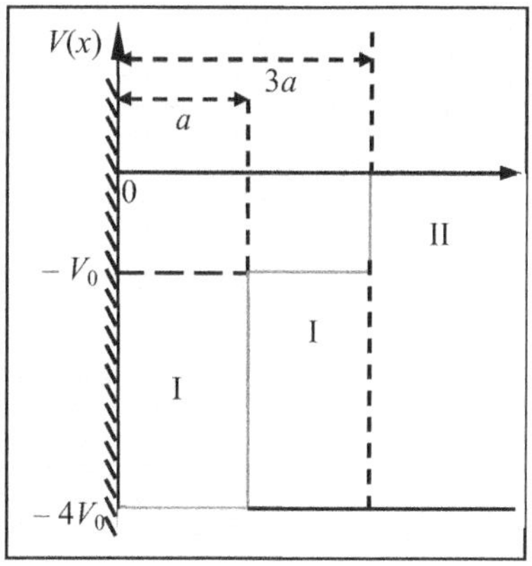

Figure 1.12. *Multistep potential*

The potential is defined as follows:

$$V(x) = \begin{cases} +\infty, & x < 0 \\ -4V_0, & 0 < x < a \\ -V_0, & a < x < 3a \\ 0, & x > 3a \end{cases}, \text{ with } V_0 > 0 \text{ and } a > 0$$

A particle of energy $E < 0$ moves from $-\infty$ to the potential step.

The profile of the potential is schematically represented in Figure 1.12.

(1) Write the stationary Schrödinger equation in zones I, II and III.

(2) The particular case for which $E = -V_0$ is studied. Let $\Psi_i(x)$ be the solution in the considered zone (i). We have:

$$k^2 = \frac{6mV_0}{\hbar^2}; q^2 = \frac{2mV_0}{\hbar^2}$$

Express the solutions $\Psi_i(x)$ in the three zones.

(3) Using the connection conditions, express the solutions $\Psi_i(x)$ as a function of x, a, and q.

(4) Then prove the relation:

$$\tan Y = -2Y - \sqrt{3}$$

where Y is a parameter to be clarified.

(5) What is the condition to obtain a bound state of energy $E = - V_0$?

(6) Use a graphical representation to find the set of solutions corresponding to a bound state of the particle of energy $E = - V_0$.

1.7.5. *Exercise 5 – Particle confined in a rectangular potential*

We study the behavior of a particle confined in an infinitely deep *rectangular potential* well (Figure 1.13).

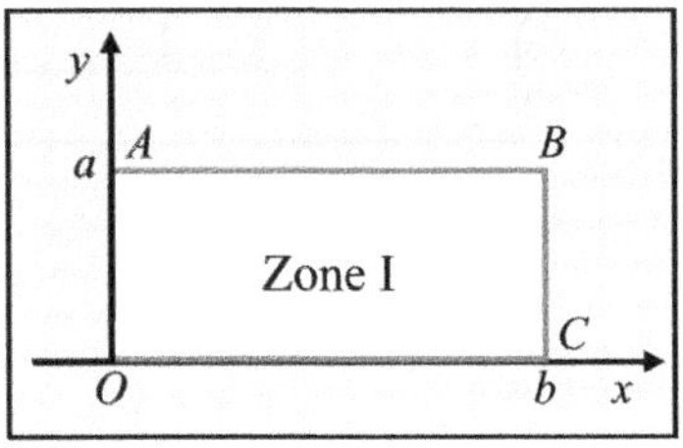

Figure 1.13. *Rectangular potential*

In zone I, the potential is marked by the points O (0, 0), A (0, a), B (b, a) and C (b, 0). This yields:

$$V(x) = \begin{cases} 0, & zone \text{ I} \\ +\infty, & elsewhere \end{cases}$$

Moreover, along the two dimensions, the total energy $E = E_x + E_y$ = constant. Let Ψ (x, y) be the wave function describing the state of the particle in zone I. $\Phi(x)$ and $\psi(y)$ designate the functions describing the state of the particle along the directions x and y, respectively.

(1) Describe the behavior of the particle from a classical perspective.

(2) Establish the differential equations satisfied by $\Phi(x)$ and $\psi(y)$.

(3) Deduce the solutions $\Phi_n(x)$ and $\psi_q(y)$ to these equations, n and q designating the quantum numbers characterizing the state of the particle along directions x and y.

(4) Find the expression of the normed wave function Ψ_{nq} (x, y). What are the minimal values of n and q?

(5) Express the density of the probability of finding the particle at point $M(x, y)$. At what point(s) is this density maximal?

(6) Prove that the total energy of the particle is quantized.

(7) Deduce the expression of the energy E' in the case of a square potential ($a = b = l$). Then show that the energy levels E' are degenerate. Specify the origin of this degeneracy.

1.7.6. *Exercise 6 – Square potential well: unbound states*

Let us consider a square potential well of width $2a$ and depth $V_0 > 0$ as shown in Figure 1.14. The potential described in this figure satisfies the following conditions:

$$V(x) = \begin{cases} 0, & x < -a \\ -V_0, & -a < x < a \\ 0, & x > a \end{cases}$$

A particle of total energy $E > 0$ comes from $-\infty$ toward the well.

The behavior of the particle upon its arrival above the well is studied.

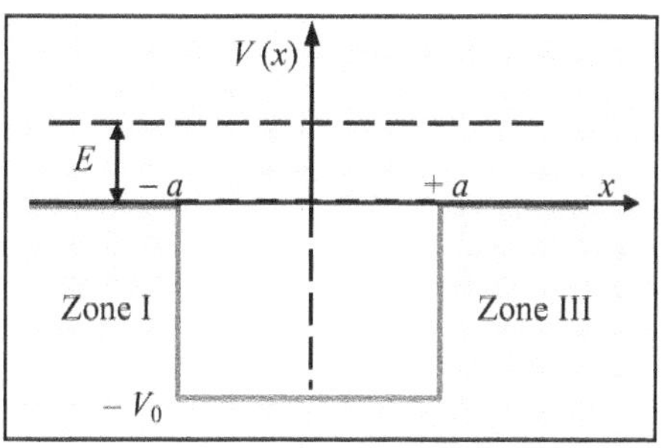

Figure 1.14. *Square potential well*

(1) How does the particle behave from a classical perspective once it arrives above the well? What is its behavior from the quantum perspective?

(2) Write the stationary Schrödinger equation in zones I, II and III. Deduce the corresponding solutions $\Phi_I(x)$, $\Phi_{II}(x)$ and $\Phi_{III}(x)$, respectively. We consider:

$k^2 = \dfrac{2mE}{\hbar^2}$ for the solutions in zones I and III.

$$\rho^2 = \frac{2m(E+V_0)}{\hbar^2} \text{ for the solution in zone II.}$$

(3) Prove that the probability of transmission from zone I to zone III can be written in the following form:

$$T = \frac{4\rho^2 k^2}{4\rho^2 k^2 + (\rho^2 - k^2)^2 \sin^2 2\rho a}$$

(4) Define and then express the probability of reflection R.

(5) In relation to the behavior of the particle, specify the predictions of quantum mechanics in comparison to classical predictions.

(6) Is the energy spectrum of the particle discrete or continuous? Justify the answer.

1.7.7. *Exercise 7 – Square potential well: bound states*

The same potential well of width $2a$ and depth $V_0 > 0$ such as that described in Figure 1.14 is considered. This time a particle is coming from $-\infty$ with a total energy $E < 0$ so that $-V_0 < E < 0$.

(1) From a classical perspective, what is the behavior of the particle since its entry in zone I (see Figure 1.14)? What is the quantum perspective?

(2) Find the wave functions $\Phi_{\text{I}}(x)$, $\Phi_{\text{II}}(x)$ and $\Phi_{\text{III}}(x)$. We consider:

$$\rho^2 = -\frac{2mE}{\hbar^2} \text{ for the wave functions in zones I and III;}$$

$$k^2 = \frac{2m(E+V_0)}{\hbar^2} \text{ for the solution in zone II.}$$

(3) Prove the relation:

$$\left(\frac{\rho - ik}{\rho + ik}\right)^2 = e^{4ika}$$

Then show that the energy is quantized.

(4) Provide a graphical solution to the above equation. Two cases are distinguished. Show that the wave functions associated with the bound states of the particle have well-defined parity.

1.7.8. *Exercise 8 – Infinitely deep rectangular potential well*

The objective is the study of the behavior of a particle in an infinitely deep potential well, the profile of which is schematically represented in Figure 1.15. The potential $V(x)$ is defined as follows:

$$V(x)=\begin{cases}0, & |x|\le a/2\\ +\infty, & |x|>a/2\end{cases}$$

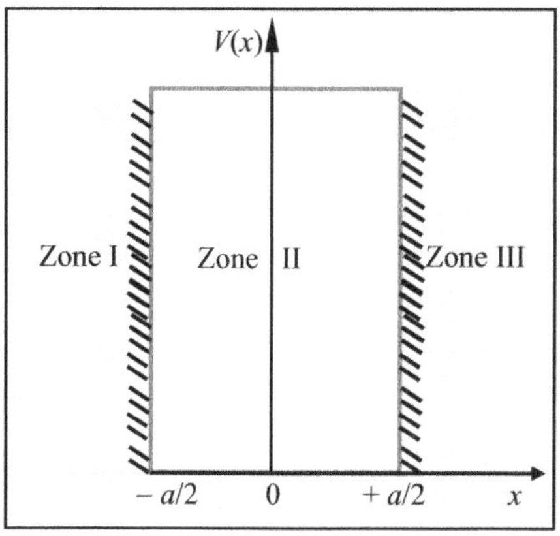

Figure 1.15. *Infinitely deep potential well*

(1) Find the wave function $\Phi(x)$ describing the state of the particle.

We consider:

$$k^2=\frac{2mE}{\hbar^2}$$

(2) What is the condition imposed on k? Then express the quantized energy of the particle.

(3) Express the even and odd wave functions describing the bound states of the particle. Establish the normed expressions of these even and odd wave functions.

(4) Provide a graphical representation of the wave functions and of the densities of probability corresponding to the ground state and to the first three excited levels of the particle.

1.7.9. *Exercise 9 – Metal assimilated to a potential well, cold emission*

In a metal, there are two bands that play a significant role in the electric conduction mechanism in metals. These are the *valence band* and the *conduction band*. In a first approximation, the N conduction electrons are considered free (in fact, each electron interacts with N – 1 other electrons and with the electric field generated by the crystal lattice).

For the model of free electrons (*perfect electron gas* model), the metal is assimilated to a *rectangular potential well* of finite depth $V_0 > 0$. Electrons are thus confined in the well (Figure 1.16). Then the potential is considered zero on the bottom of the well and changes at the walls, jumping from 0 to V_0. At low temperature, no electron can be emitted by the metal. Nevertheless, if the metal is brought to quite a high temperature (above 1,000°C), the thermal agitation motion becomes so intense that electrons with sufficient energy manage to overcome the *potential barrier* of height V_0 and escape from the metal: this phenomenon is known as the *thermoelectric effect*. Is it then possible to observe a cold metal emitting electrons?

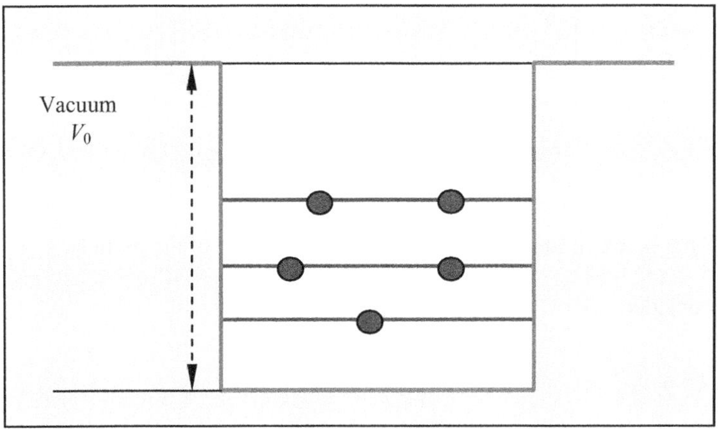

Figure 1.16. *Metal assimilated to a rectangular potential well*

From an experimental point of view, it can be noted that when a strong electric field (approximately 10^6 V/cm) is applied, normally at the surface of a metal, the latter emits electrons: it is the *cold emission* phenomenon that is studied in this exercise. When an electric field is zero, the potential energy is represented by a step $AOBC$ of origin O located at the surface of the metal. The potential energy can therefore be considered zero inside the metal and equal to a constant K outside the metal (Figure 1.17).

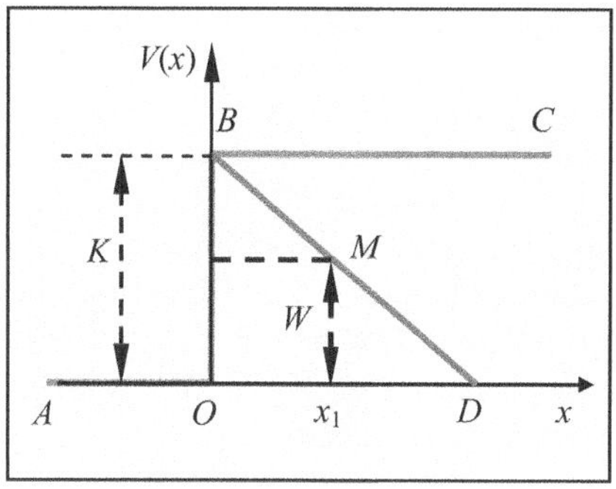

Figure 1.17. *Potential profile in a metal assimilated to a rectangular potential well*

If an electric field of intensity E is applied, this field does not penetrate the metal, and so the potential is still zero. Outside the metal, the potential energy varies as a function of x and decreases according to BD. Hence, between the metal and the vacuum there is a potential barrier OBD. An electron of energy W at the point M of abscissa x_1 (thickness of the barrier at point M) is *tunneling* out of the metal.

Transparency T of the barrier is given by the expression [SIV 86, SAK 12]:

$$T = \exp - 2\int_0^{x_1} \sqrt{\frac{2m[V(x) - W(x)]}{\hbar}}\, dx$$

(1) Find the expression of potential energy $V(x)$ for $x > 0$.

(2) Express x_1 as a function of K, E, W and e (elementary charge).

(3) Prove that the probability of transmission of the barrier can be written in the following form ($\phi = K - W$):

$$T = \exp - E_0 / E$$

where E_0 is a constant whose expression will be specified.

Calculate E_0 and T. Draw a conclusion.

Given data.

$e = 1.6 \times 10^{-19}$ C; $m = 9.1\times 10^{-31}$ kg; $E = 4 \times 10^{7}$ V·cm^{-1}

$\hbar = 1.05 \times 10^{-34}$ J · s; $\phi = 5$ eV

1.7.10. *Exercise 10 – Ground state energy of the harmonic oscillator*

Let us consider a one-dimensional classical harmonic oscillator. It is constituted of a particle of mass m. During the motion, the position of the oscillator is given by its abscissa x with respect to a point O chosen as origin of space. At any instant, the oscillator is subjected to an *opposing spring force* $F = -kx$, where k is a positive constant known as the *coefficient of elasticity*.

(1) Express the elastic potential energy $V(x)$ of the classical oscillator.

(2) Prove that the classical oscillator is a conservative system.

(3) Let us now study the behavior of a quantum harmonic oscillator of potential energy $V(x)$.

(3.1) Prove that the stationary Schrödinger equation describing the evolution of the quantum harmonic oscillator can be written in the following form:

$$-\frac{d^2\Phi(q)}{dq^2} + q^2\Phi(q) = \alpha\, \Phi(q)$$

where q and α are dimensionless quantities to be specified.

(3.2) For a certain value of α, the ground state wave function has the form $\Phi_0(q) = \exp(\beta q^2)$, where β is a constant. Using the previous equation, prove the relation:

$$(1 - 4\beta^2)q^2 - 2\beta = \alpha$$

(3.3) Deduce from this equation the possible values of β. What value should be retained? Justify the answer.

(3.4) Find the expression of the ground state energy E_0 of the quantum harmonic oscillator.

1.7.11. *Exercise 11 – Quantized energy of the harmonic oscillator*

Let us consider a one-dimensional quantum harmonic oscillator of energy E, angular frequency ω and potential energy $V(x)$. The dimensionless quantities are:

$$q=\sqrt{\frac{m\omega}{\hbar}}x\,;\ \varepsilon=\frac{2E}{\hbar\omega}$$

(1) Prove that the Schrödinger equation for the stationary states of the harmonic oscillator can be written in the following form:

$$\frac{d^2\Phi(q)}{dq^2}+\left(\varepsilon-q^2\right)\Phi(q)=0$$

(2) The solution to this equation has the form:

$$\Phi(q)=Au(q)e^{-q^2/2}$$

In this expression, A is a constant to be determined by the normalization condition.

The function $u\,(q)$ is a complete series of powers of q given by the expression:

$$u(q)=\sum_{k=0}^{\infty}a_k\,q^k$$

Prove that $u\,(q)$ satisfies the following differential equation:

$$\frac{d^2u(q)}{dq^2}-2q\frac{du(q)}{dq}+(\varepsilon-1)\,u(q)=0$$

(3) Express the *recurrence relation* satisfied by the coefficients of $u\,(q)$.

(4) Using the *cut-off condition*, prove that the energy of the harmonic oscillator is quantized. Deduce the value of the energy E_0 of the ground state.

(5) Plot the curve of the variation of potential energy $V(x)$. Draw on this curve the ground level of the studied quantum harmonic oscillator, as well as the first four excited levels. Comment on this curve first from a classical perspective and then from a quantum perspective.

(6) What is the energy of the studied harmonic oscillator from both a classical and a quantum point of view?

(7) Making the *classical oscillator–quantum oscillator analogy*, decide if the existence of energy E_0 can be justified from the classical point of view.

1.7.12. *Exercise 12 – HCl molecule assimilated to a linear oscillator*

A very simple particular case of a quantum harmonic oscillator is the model of the hydrogen chloride molecule HCl assimilated to an oscillating dipole.

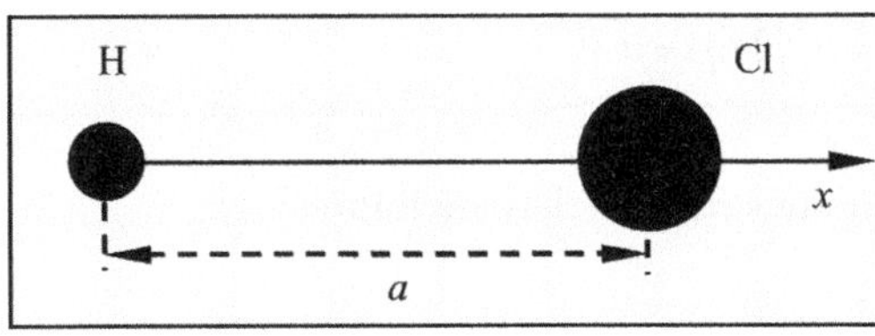

Figure 1.18. *HCl molecule assimilated to an oscillating dipole*

Indeed, since chlorine is more electronegative than hydrogen, the electron doublet is not equidistant from the centers of inertia of H and Cl atoms. The doublet is slightly attracted by the chlorine, which induces a polarization of the molecule and each of the bound atoms also carries a partial charge –δ for Cl and +δ for H. The mean distance between the hydrogen atom and the chlorine atom is denoted by a (Figure 1.18). The potential energy of the dipole thus constituted has the following form:

$$V(x) = \frac{1}{2}k(x-a)^2$$

Moreover, the wave function $\Phi_0(x)$ of the ground state and $\Phi_1(x)$ of the first excited state are given by the expressions:

$$\Phi_0(x) = A_0 e^{-\alpha^2(x-a)^2/2};\ \Phi_1(x) = A_1(x-a)e^{-\beta^2(x-a)^2/2}$$

In these expressions, A_0 and A_1 are normalization constants and α and β are strictly positive constants.

(1) Write the Schrödinger equation of the vibration stationary states of the hydrogen chloride molecule. Deduce from it the relation between α and β.

(2) Find the expressions of energies E_0 and E_1 of the respective ground state and first excited state of the HCl molecule.

(3) What are the values of constants A_0 and A_1? Deduce the expressions of the normed wave functions of the ground state and of the first excited state.

Given data.

For the family of integrals of the type:

$$I_p = \int_0^{\infty} x^p e^{-\rho x^2} dx$$

(where ρ is a strictly positive constant), the recurrence relation can be written as:

$$I_p = \frac{p-1}{2\rho} I_{p-2};\ I_0 = \frac{1}{2}\sqrt{\frac{\pi}{\rho}};\ I_1 = \frac{1}{2\rho}$$

1.7.13. *Exercise 13 – Quantized energy of hydrogen-like systems*

The objective here is to determine the expression of the quantized energy of hydrogen-like systems. For such systems which are in a stationary state, we consider the wave functions with spherical symmetry that depend only on the radial variable r.

The Schrödinger equation describing the evolution of the radial function is of a similar type to equation [1.144]. The following changes are made:

$\sigma = -\varepsilon^2$ and $2\rho = \delta$

The parameters σ and ρ are given by relations [1.143]. Consequently, equation [1.144] has the form:

$$\frac{d^2\Phi(r)}{dr^2} + \frac{2}{r}\frac{d\Phi(r)}{dr} + \left(\frac{\delta}{r} - \varepsilon^2\right)\Phi(r) = 0 \qquad \text{(equation 1)}$$

In this equation, the variation of the wave function with r is given by the expression:

$$\Phi(r) = \frac{\chi(r)}{r} e^{-\varepsilon r} \qquad \text{(equation 2)}$$

The function $\chi(\mathrm{r})$ in equation (2) is written as a complete series:

$$\chi(r) = \sum_{k=\nu}^{\infty} a_k \, r^k \qquad \text{(equation 3)}$$

where ν is a positive integer to be determined.

(1) Using equation (1), establish the differential equation verified by the function $\chi(r)$.

(2.1) Using equation (3), prove the following relations:

$$\begin{cases} \nu(\nu-1) & = 0 \\ k(k+1)a_{k+1} - 2\varepsilon k \, a_k + \delta \, a_k & = 0 \end{cases} \qquad \text{(equations 4)}$$

(2.2) Deduce from equations (4) the possible values of ν. What values should be retained? Why?

(3) Express the ratio a_{k+1}/a_k. What is the asymptotic behavior of this ratio to infinity?

(4) Compare the behavior of the ratio a_{k+1}/a_k to infinity to that of the complete series expansion of the function $e^{2\varepsilon r}$. Draw a conclusion.

(5) Using the cut-off condition, express the quantized energy of hydrogen-like systems.

1.7.14. *Exercise 14 – Line integral of the probability current density vector, Bohr's magneton*

At various points in Chapter 3 of Volume 1, we expressed the energy gaps between the fine structure levels as a function of Bohr's magneton μ_B (see, for example, the exercise in section 3.7.15). The objective here is to establish the expression of μ_B based on the probability current density.

In classical electrodynamics, the expression of the magnetic moment of a circular current features the intensity of the circular current and the area around which it flows. Nevertheless, in quantum mechanics, which rejects the notion of circular or elliptical orbit, the situation is described in terms of density of probability of the presence of the electron. This approach takes into account the *mean density of the electrical charge* $e\Psi^*\Psi$ present throughout the space (charge is distributed in space and is not confined to the line assimilated to a circular or elliptical loop).

The mean value of the current is thus the product of the elementary charge e and the probability current density $\vec{J}$ provided by expression [1.48], which is recalled as follows:

$$\vec{J}(\vec{r},t)=\frac{\hbar}{2mi}\left[\Psi^*\vec{\nabla}\Psi-\Psi\vec{\nabla}\Psi^*\right]$$

In this expression, the wave function depends on spherical coordinates (r, θ, φ). Therefore, $\Psi = \Psi(r, \theta, \varphi) = R(r) \times \Theta(\theta) \times \Phi(\varphi)$. The radial part $R(r)$ and the angular part $\Theta(\theta)$ of the wave function are real. The angular part $\Phi(\varphi) = \exp(i m_\ell \varphi)$ is complex. Figure 1.19 shows the flow of the tube of volume current (*equivalent of the circular loop in classical theory)*, the component J_φ of the probability current density vector and the area $d\sigma$ of the cross-section of the current tube.

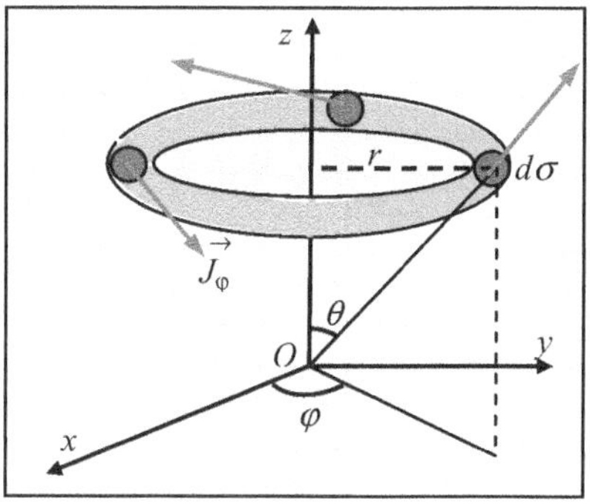

Figure 1.19. *Tube of volume current*

(1) Specify the values of the components J_r and J_θ of the probability current density vector. Prove that J_φ can be written in the form:

$$J_\varphi = \frac{\hbar}{m} m_\ell \times \frac{\Psi^*\Psi}{r\sin\theta}$$

(2) Express the intensity dI_φ of the current through $d\sigma$ (which is the *flux of the probability current density vector* through the elementary surface $d\sigma$).

(3) Prove that the *magnetic moment* dM_ℓ verifies the relation:

$$dM_\ell = \frac{e\hbar}{2m} m_\ell \Psi^*\Psi\, d\tau$$

where $d\tau$ designates the elementary volume of the current tube of cross-section $d\sigma$.

(4) Express the orbital magnetic moment of the electron. Deduce the expression of Bohr's magneton.

Given data. Components of the gradient in spherical coordinates:

$$(\vec{\nabla}\Psi)_r = \frac{\partial \Psi}{\partial r}\,;\, (\vec{\nabla}\Psi)_\theta = \frac{1}{r}\frac{\partial \Psi}{\partial \theta}\,;(\vec{\nabla}\Psi)_\varphi = \frac{1}{r\sin\theta}\frac{\partial \Psi}{\partial \varphi}$$

1.7.15. *Exercise 15 – Schrödinger's equation in the presence of a magnetic field, Zeeman–Lorentz triplet*

Schrödinger's equation offers a simple way to find the previous Zeeman–Lorentz triplet (result [3.188]). This proves once more the broad range of applications of this equation for the correct interpretation of various physical phenomena such as the *Zeeman effect*, which is the focus of this exercise. In the absence of a field of external forces, the Schrödinger equation describing the evolution of the state of a free particle is written as follows:

$$\Delta\Psi + \frac{2m}{\hbar^2}(E - V)\Psi = 0$$

In the presence of a magnetic field, an additional term appears [CHP 78]:

$$-i\frac{2e}{\hbar}\vec{A}.\vec{\nabla}\Psi \qquad \text{(equation 1)}$$

This term is due to the action of the magnetic field on the particle and $\vec{A}$ corresponds to the *vector potential* and satisfies the equation $\vec{B} = \vec{\nabla} \wedge \vec{A}$.

Schrödinger's equation in the presence of a magnetic field is in this case:

$$\Delta\Psi - i\frac{2e}{\hbar}\vec{A}.\vec{\nabla}\Psi + \frac{2m}{\hbar^2}(E - V)\Psi = 0 \qquad \text{(equation 2)}$$

Let us consider a hydrogen-like system (fixed nucleus) subjected to a uniform magnetic field of arbitrary direction Oz. The wave function of the hydrogen-like system considered is $\Psi\ (r,\theta,\varphi) = \mathrm{R}\ (r) \times \Theta\ (\theta) \times \Phi\ (\varphi)$. It is worth recalling that functions $\mathrm{R}\ (r)$ and $\Theta\ (\theta)$ are real and $\Phi\ (\varphi) = \exp\ (i\, m_\ell\ \varphi)$.

In the Zeeman effect, only the function $\Phi\ (\varphi)$ containing the orbital magnetic quantum number plays an essential role.

(1) Passing to spherical coordinates and assuming that variables r and θ are constant, prove the relation:

$$\frac{\partial \Psi}{\partial \varphi} = x\frac{\partial \Psi}{\partial y} - y\frac{\partial \Psi}{\partial x}$$

(2) Prove that the applied magnetic field is deduced from the vector potential, the components of which are:

$$A_x = -\frac{1}{2}B\,y\,;\; A_y = \frac{1}{2}B\,x\,;\, A_z = 0$$

(3) Prove the relation:

$$-i\frac{2e}{\hbar}\vec{A}.\vec{\nabla}\Psi = \frac{e}{\hbar}Bm_\ell\Psi$$

(4) Write the Schrödinger equation describing the evolution of the state of the electron along the direction of the magnetic field using equation (2). Prove that it can be written in the form:

$$\Delta\Psi + \frac{2m}{\hbar^2}(E' - V)\Psi = 0 \qquad \text{(equation 3)}$$

In this equation, E' is an energy that will be defined and expressed.

(5) For a hydrogen-like system whose potential energy is $V(r) = -Ze^2/r$ and total energy is E, Schrödinger's equation can be written as follows:

$$\Delta\Psi + \frac{2m}{\hbar^2}(E - V)\Psi = 0$$

Knowing that solving this equation yields discrete values E_n of the energy E (see the exercise in section 1.7.13), prove without solving equation (3) that energy E' is quantized. Then find the *Zeeman–Lorentz triplet*.

Given data. $x = r\sin\theta\cos\varphi$, $y = r\sin\theta\sin\varphi$, $z = r\cos\theta$.

1.7.16. *Exercise 16 – Deduction of the stationary Schrödinger equation from the De Broglie relation*

The wave function describing the evolution of a system makes it possible to calculate the probability of finding the system at a point $\vec{r}$ in space. This function can also be used to calculate the mean values of the physical quantities characterizing the system, such as energy and linear momentum. The problem is then knowing how to find such a wave function.

Obviously, the De Broglie plane wave cannot be used, as its temporal and spatial uniformity is incompatible with the space localization of the system. Schrödinger holds the merit of postulating in 1926 the fundamental equation of quantum mechanics describing the evolution of any wave function. There is no logical approach to proving this equation (since it originates in a postulate). Its validity is confirmed by the significant physical consequences following from it. There is, nevertheless, a purely inductive way to find the Schrödinger equation.

The objective of this exercise is to find this equation from the De Broglie relation. For this purpose, let us consider a one-dimensional problem. The generalized coordinate is designated by q, $q = x, y$ or z.

Let $\Psi(q, t)$ be an electromagnetic wave propagating in vacuum. It satisfies the general differential equation:

$$\frac{\partial^2 \Psi}{\partial q^2} = \frac{1}{c^2}\frac{\partial^2 \Psi}{\partial t^2} \qquad \text{(equation 1)}$$

Schrödinger's equation being verified for any quantum state, we consider the particular case of a stationary state for which the wave function solution to the propagation (equation (1)) has the form:

$$\Psi(q,t) = \Phi(q) \times \exp(i\omega t)$$

As chosen, the stationary wave function $\Psi(q, \mathrm{t})$ oscillates in time with angular frequency ω, its amplitude varying with the generalized coordinate q.

(1) Express equation (1) as a function of the wavelength and of the frequency with which the electromagnetic wave $\Psi(q, t)$ oscillates in time.

(2) Use De Broglie's relation to prove:

$$\frac{\partial^2 \Phi(q)}{\partial q^2} = -4\pi^2 \frac{p^2}{h^2}\Phi(q)$$

(3) Express the energy E of a conservative system subjected to potential $V(q)$.

(4) Find the one-dimensional stationary Schrödinger equation.

(5) Make a generalization of the previous result to three dimensions.

1.8. Solutions

1.8.1. *Solution 1 – Probability current density*

(1) *Hamiltonian expression, operator actions*

The Hamiltonian H of the system and the equations resulting from the action of operators Q and $V(Q, \mathrm{t})$ on the wave function $\Psi(q, t)$ are given by the relations:

$$H = -\frac{\hbar^2}{2m}\Delta_q + V(Q,t) \quad [1.149]$$

$$Q\Psi(q, t) = q\Psi(q, t);\ V(Q, t)\Psi(q, t) = V(q, t)\Psi(q, t) \quad [1.150]$$

(2) *Proof*

The one-dimensional probability density is written according to [1.48]:

$$\rho(q,t) = |\Psi(q,t)|^2 = \Psi(q,t)\Psi^*(q,t) \quad [1.151]$$

For simplicity purposes, variables q and t in relation [1.151] are omitted. We express the first derivative with respect to time of the probability density. We obtain:

$$\frac{\partial\rho}{\partial t} = \frac{\partial\Psi^*}{\partial t}\Psi + \Psi^*\frac{\partial\Psi}{\partial t} = -\frac{1}{i\hbar}[(H\Psi)^*]\Psi + \frac{1}{i\hbar}\Psi^*[H\Psi] \quad [1.152]$$

In [1.152], we substitute H by its expression [1.149], hence:

$$\frac{\partial\rho}{\partial t} = -\frac{1}{i\hbar}\left[-\frac{\hbar^2}{2m}\Delta_q\Psi^* + V(q,t)\Psi^*\right]\Psi + \frac{1}{i\hbar}\Psi^*\left[-\frac{\hbar^2}{2m}\Delta_q\Psi + V(q,t)\Psi\right]$$

This is:

$$\frac{\partial\rho}{\partial t} = \frac{\hbar}{2im}\left[\Psi(\Delta_q\Psi^*) - (\Delta_q\Psi)\Psi^*\right] \quad [1.153]$$

Moreover, the one-dimensional expression of the linear momentum operator is written according to [1.14]:

$$P_q = -i\hbar\Delta_q = -i\hbar\frac{\partial}{\partial q} \quad [1.154]$$

Using [1.154], relation [1.153] is written as:

$$\frac{\partial \rho}{\partial t}=\frac{\hbar}{2im}\left[\Psi\left(\frac{\partial^2\Psi^*}{\partial q^2}\right)-\left(\frac{\partial^2\Psi}{\partial q^2}\right)\Psi^*\right] \tag{1.155}$$

(3) *Proof*

Let us consider the following system of equations:

$$\begin{cases}\dfrac{\partial}{\partial q}\left(\Psi\dfrac{\partial\Psi^*}{\partial q}\right)=\dfrac{\partial\Psi}{\partial q}\dfrac{\partial\Psi^*}{\partial q}+\Psi\left(\dfrac{\partial^2\Psi^*}{\partial q^2}\right)\\[2ex] \dfrac{\partial}{\partial q}\left(\Psi^*\dfrac{\partial\Psi}{\partial q}\right)=\dfrac{\partial\Psi^*}{\partial q}\dfrac{\partial\Psi}{\partial q}+\Psi^*\left(\dfrac{\partial^2\Psi}{\partial q^2}\right)\end{cases} \tag{1.156}$$

Subtracting one relation from the other, member by member, we have:

$$\Psi\left(\frac{\partial^2\Psi^*}{\partial q^2}\right)-\left(\frac{\partial^2\Psi}{\partial q^2}\right)\Psi^*=\frac{\partial}{\partial q}\left(\Psi\frac{\partial\Psi^*}{\partial q}\right)-\frac{\partial}{\partial q}\left(\Psi^*\frac{\partial\Psi}{\partial q}\right)$$

Hence:

$$\Psi\left(\frac{\partial^2\Psi^*}{\partial q^2}\right)-\left(\frac{\partial^2\Psi}{\partial q^2}\right)\Psi^*=\frac{\partial}{\partial q}\left(\Psi\frac{\partial\Psi^*}{\partial q}-\frac{\partial\Psi}{\partial q}\Psi^*\right)$$

which then leads to:

$$\Psi\left(\frac{\partial^2\Psi^*}{\partial q^2}\right)-\left(\frac{\partial^2\Psi}{\partial q^2}\right)\Psi^*=-\frac{\partial}{\partial q}\left(\Psi^*\frac{\partial\Psi}{\partial q}-\frac{\partial\Psi^*}{\partial q}\Psi\right) \tag{1.157}$$

Using [1.157], equation [1.155] can be written in the following form:

$$\frac{\partial \rho}{\partial t}=-\frac{\hbar}{2im}\frac{\partial}{\partial q}\left(\Psi^*\frac{\partial\Psi}{\partial q}-\frac{\partial\Psi^*}{\partial q}\Psi\right) \tag{1.158}$$

(4) *Probability conservation*

Probability conservation is reflected by an analogous continuity equation [1.42]. Taking [1.158] into account, the *probability current density* $J_q = J(q, t)$ is written as:

$$J_q = \frac{\hbar}{2im}\left(\Psi^* \frac{\partial \Psi}{\partial q} - \frac{\partial \Psi^*}{\partial q}\Psi\right) \qquad [1.159]$$

Highlighting variables q and t, the conservation of probability can finally be written as:

$$\frac{\partial \rho(q,t)}{\partial t} + \frac{\partial J(q,t)}{\partial q} = 0 \qquad [1.160]$$

1.8.2. *Solution 2 – Heisenberg's spatial uncertainty relations*

Given data:

$$\langle x \rangle = 0 \; ; \langle p \rangle = 0 \qquad [1.161]$$

$$\int \left| ux\Psi + w\frac{d\Psi}{dx} \right|^2 dx \geq 0 ; \left(\Psi^* \frac{d\Psi}{dx}\right)\Bigg|_{-\infty}^{+\infty} = 0 \qquad [1.162]$$

(1) *Mean value, root mean square deviation*

Let us consider the one-dimensional wave function $\Psi(x, t)$. The mean value $\langle x \rangle$ is defined by the relation:

$$\langle x \rangle = \int_{-\infty}^{+\infty} x \left| \Psi(x,t) \right|^2 dx \qquad [1.163]$$

Moreover, the root mean square deviations are given by the relations:

$$\Delta x = \sqrt{\left\langle x^2 \right\rangle - \left\langle x \right\rangle^2} \; ; \Delta p = \sqrt{\left\langle p_x^2 \right\rangle - \left\langle p_x \right\rangle^2} \qquad [1.164]$$

Taking [1.161] into account, the following relations can be deduced from [1.164]:

$$\Delta x = \sqrt{\left\langle x^2 \right\rangle} \; ; \Delta p_x = \sqrt{\left\langle p_x^2 \right\rangle} \qquad [1.165]$$

(2) *Proof, sign of A*

We consider:

$$A=\int x^2\Psi\Psi^*dx\,;\; B=-\int x\frac{d}{dx}(\Psi\Psi^*)dx\,;\; C=\int\frac{d\Psi^*}{dx}\frac{d\Psi}{dx}dx \qquad [1.166]$$

Let us find the expressions under the integral sign of inequality [1.162]. We have:

$$\left|ux\Psi+w\frac{d\Psi}{dx}\right|^2=\left(ux\Psi+w\frac{d\Psi}{dx}\right)\times\left(ux\Psi+w\frac{d\Psi}{dx}\right)^*$$

Knowing that u and w are real variables, we have:

$$\left|ux\Psi+w\frac{d\Psi}{dx}\right|^2=u^2x^2\Psi\Psi^*+uwx\left(\Psi^*\frac{d\Psi}{dx}+\Psi\frac{d\Psi^*}{dx}\right)+w^2\frac{d\Psi^*}{dx}\frac{d\Psi}{dx}$$

Hence:

$$\left|ux\Psi+w\frac{d\Psi}{dx}\right|^2=u^2x^2\Psi\Psi^*+uwx\frac{d}{dx}(\Psi^*\Psi)+w^2\frac{d\Psi^*}{dx}\frac{d\Psi}{dx} \qquad [1.167]$$

Integrating [1.167], we have:

$$\int\left|ux\Psi+w\frac{d\Psi}{dx}\right|^2dx=u^2\int x^2\Psi\Psi^*\,dx+uw\int x\frac{d}{dx}(\Psi^*\Psi)\,dx+w^2\int\frac{d\Psi^*}{dx}\frac{d\Psi}{dx}\,dx$$

Or taking inequality [1.167] into account, we have:

$$Au^2-uwB+Cw^2\geq 0 \qquad [1.168]$$

Moreover, considering the expression of A according to [1.166], we see that:

$$A=\langle x^2\rangle=\int_{-\infty}^{+\infty}x^2\,|\Psi|^2dx\geq 0 \qquad [1.169]$$

(3) *Expressions of A and B, inequality verified by the product AC*

– *Expression of A*

Taking [1.165] and [1.169] into account, we get:

$$A=(\Delta x)^2 \qquad [1.170]$$

– *Expression of B*

To find the expression of B, let us integrate by parts. We have:

$$B = -\int_{-\infty}^{+\infty} x\frac{d}{dx}(\Psi\Psi^*) = -\left(x\Psi\Psi^*\right)\Big|_{-\infty}^{+\infty} + \int_{-\infty}^{+\infty}(\Psi\Psi^*)dx \qquad [1.171]$$

The wave function Ψ being square-summable, $|\Psi|^2 \to 0$ when $x \to \pm\infty$. Consequently:

$$\left(x|\Psi|^2\right)\Big|_{-\infty}^{+\infty} = 0$$

Knowing that the wave function Ψ is normed, relation [1.171] then yields:

$$B = \int_{-\infty}^{+\infty} |\Psi|^2 dx = 1 \qquad [1.172]$$

– *Inequality verified by the product AC*

Integrating by parts as previously, we have:

$$C = \int_{-\infty}^{+\infty} \frac{d\Psi^*}{dx}\frac{d\Psi}{dx}dx = \left(\Psi^*\frac{d\Psi}{dx}\right)\Big|_{-\infty}^{+\infty} - \int_{-\infty}^{+\infty}\Psi^*\frac{d^2\Psi}{dx^2}dx$$

Taking [1.162] into account, we get:

$$C = -\int_{-\infty}^{+\infty}\Psi^*\frac{d^2\Psi}{dx^2}dx \qquad [1.173]$$

Let us now determine the mean value $\langle p_x^2\rangle$.

Operator

$$p_x = -i\hbar\frac{d}{dx}$$

We have:

$$\langle p_x^2\rangle = \int_{-\infty}^{+\infty}\Psi^* p_x^2\Psi dx = -\hbar^2\int_{-\infty}^{+\infty}\Psi^*\frac{d^2\Psi}{dx^2}dx$$

Using this result, we see that, according to [1.173] and [1.165], we have:

$$C = \frac{\langle p_x^2 \rangle}{\hbar^2} \Rightarrow C = \frac{(\Delta p_x)^2}{\hbar^2} \qquad [1.174]$$

(4) *Expression of the product AC*

Let us consider the second-degree equation in u according to [1.168]. We obtain:

$$Au^2 - uwB + Cw^2 = 0 \Rightarrow \Delta = w^2B^2 - 4ACw^2$$

Knowing that $A \geq 0$ according to [1.169] and $B = 1$ according to [1.172], this means:

$$B^2 - 4AC \leq 0 \Rightarrow 4AC \geq 1 \qquad [1.175]$$

(5) *Heisenberg uncertainty relations*

Considering results [1.170] and [1.174], inequality [1.175] is written as:

$$4(\Delta x)^2(\Delta p_x)^2 \geq \hbar^2 \Rightarrow \Delta x \Delta p_x \geq \frac{\hbar}{2} \qquad [1.176]$$

Applying circular permutation, *Heisenberg spatial uncertainty relations* are written in the following form:

$$\Delta x \Delta p_x \geq \frac{\hbar}{2}; \Delta y \Delta p_y \geq \frac{\hbar}{2}; \Delta z \Delta p_z \geq \frac{\hbar}{2} \qquad [1.177]$$

1.8.3. *Solution 3 – Finite-depth potential step*

(1) *First case: $E > V_1$*

(1.1) *Nature of the states of the particle, Schrödinger equation*

Potential depends only on variable x in zones I and II. The Hamiltonian is therefore time independent: consequently, the states of the particle are stationary states. The Schrödinger equation is written for a zone i under consideration:

$$\frac{d^2\Phi(x)}{dx^2} + \frac{2m}{\hbar^2}[E - V(x)]\Phi(x) = 0 \qquad [1.178]$$

In zones I ($x < a$) and II ($x > a$), the stationary Schrödinger equation is written respectively as:

$$\frac{d^2\Phi_I(x)}{dx^2} + \frac{2m}{\hbar^2}[E + V_2]\Phi_I(x) = 0 \qquad [1.179]$$

$$\frac{d^2\Phi_{II}(x)}{dx^2} + \frac{2m}{\hbar^2}[E - V_1]\Phi_{II}(x) = 0 \qquad [1.180]$$

(1.2) *Solutions in zones I and II*

We consider:

$$q^2 = \frac{2m(E + V_2)}{\hbar^2}; \rho^2 = \frac{2m(E - V_1)}{\hbar^2} \qquad [1.181]$$

Equations [1.179] and [1.180] admit the following solutions:

$$\begin{cases} \Phi_I(x) = A\,e^{iq(x-a)} + B\,e^{-iq(x-a)} \\ \\ \Phi_{II}(x) = C\,e^{i\rho(x-a)} + D\,e^{-i\rho(x-a)} \end{cases} \qquad [1.182]$$

Knowing there is no reflected wave in zone II, then $D = 0$. This finally leads to:

$$\begin{cases} \Phi_I(x) = A\,e^{iq(x-a)} + B\,e^{-iq(x-a)} \\ \\ \Phi_{II}(x) = C\,e^{i\rho(x-a)} \end{cases} \qquad [1.183]$$

(1.3) *Expressions of transmission and reflection probabilities*

Let us express the boundary conditions for the wave function in $x = a$:

$$\Phi_I(0) = \Phi_{II}(0)$$

$$\Phi'_I(0) = \Phi'_{II}(0) \qquad [1.184]$$

$$\Phi'_i(x - a) = d\Phi_i(x - a)/dx$$

Applying these boundary conditions to equations [1.183], we find:

$$\begin{cases} A+B=C \\ q(A-B)=\rho C \end{cases} \Rightarrow \begin{cases} q(A+B)=qC \\ q(A-B)=\rho C \end{cases} \qquad [1.185]$$

We express the coefficients B and C as a function of A (*coefficient assigned to the incident wave*). Making the sum and then the difference of the two equations [1.185], we find:

$$C=\frac{2q}{q+\rho}A\,;\,B=\frac{q-\rho}{q+\rho}A \qquad [1.186]$$

Using [1.186], the probabilities of reflection and transmission are written as:

$$R=\frac{|B|^2}{|A|^2}\Rightarrow R=\frac{(q-\rho)^2}{(q+\rho)^2} \qquad [1.187]$$

$$T=\frac{\rho}{q}\times\frac{|C|^2}{|A|^2}\Rightarrow T=\frac{4\rho q}{(q+\rho)^2} \qquad [1.188]$$

(1.4) *Classical and quantum predictions*

Considering [1.187] and [1.188], we have:

$$R+T=\frac{(q-\rho)^2}{(q+\rho)^2}+\frac{4\rho q}{(q+\rho)^2}=1$$

CONCLUSION.– $R + T = 1$: The particle is either reflected or transmitted, contrary to the predictions of classical mechanics, according to which the particle can in no way be reflected since $E > V_1$.

(2) *Second case: $-V_2 < E < V_1$*

(2.1) *Schrödinger equation, solutions*

The stationary Schrödinger equation in zones I (x < a) and II (x > a) is given by [1.179] and [1.180], respectively. To determine the solutions $\Phi_I(x)$ and $\Phi_{II}(x)$, let us consider:

$$q^2 = \frac{2m(E+V_2)}{\hbar^2}; \sigma^2 = \frac{2m(V_1 - E)}{\hbar^2} = -\rho^2 \Rightarrow \rho = i\sigma \qquad [1.189]$$

Equations [1.179] and [1.180] admit the following solutions (replacing ρ by $i\sigma$):

$$\begin{cases} \Phi_{\mathrm{I}}(x) = Ae^{iq(x-a)} + Be^{-iq(x-a)} \\ \\ \Phi_{\mathrm{II}}(x) = Ce^{-\sigma(x-a)} + De^{\sigma(x-a)} \end{cases} \qquad [1.190]$$

Knowing that a wave function must be square-summable, the coefficient $D = 0$. Therefore, we finally have:

$$\begin{cases} \Phi_{\mathrm{I}}(x) = Ae^{iq(x-a)} + Be^{-iq(x-a)} \\ \\ \Phi_{\mathrm{II}}(x) = Ce^{-\sigma(x-a)} \end{cases} \qquad [1.191]$$

(2.2) *Expressions of T and R*

Let us express the boundary conditions [1.184] in $x = a$:

$$\begin{cases} A + B = C \\ iq(A - B) = -\sigma C \end{cases} \Rightarrow \begin{cases} iq(A + B) = iqC \\ iq(A - B) = -\sigma C \end{cases} \qquad [1.192]$$

Using [1.192], we obtain:

$$\begin{cases} C = \dfrac{2q}{q + i\sigma} A \\ B = \dfrac{q - i\sigma}{q + i\sigma} A \end{cases} \Rightarrow R = \frac{|B|^2}{|A|^2} = \frac{|q - i\sigma|^2}{|q + i\sigma|^2} = 1 \qquad [1.193]$$

Since $R = 1$, then $T = 0$.

CONCLUSION.– Reflection is total, according to the predictions of classical mechanics. Nevertheless, the wave in zone II is not zero, as shown by the second equation [1.191].

1.8.4. Solution 4 – Multistep potential

(1) *Schrödinger's equation, solutions*

Schrödinger's equation for stationary states in zones I, II and III is written as:

$$\frac{d^2\Phi_{\text{I}}(x)}{dx^2}+\frac{2m}{\hbar^2}(E+4V_0)\Phi_{\text{I}}(x)=0 \quad [1.194]$$

$$\frac{d^2\Phi_{\text{II}}(x)}{dx^2}+\frac{2m}{\hbar^2}(E+V_0)\Phi_{\text{II}}(x)=0 \quad [1.195]$$

$$\frac{d^2\Phi_{\text{III}}(x)}{dx^2}+\frac{2mE}{\hbar^2}\Phi_{\text{III}}(x)=0 \quad [1.196]$$

We consider:

$$k^2=\frac{2m(E+4V_0)}{\hbar^2};\; q^2=\frac{2mE}{\hbar^2} \quad [1.197]$$

(2) *Expression of the solutions in the three zones*

Zone I

The solution to equation [1.194] has the form:

$$\Phi_{\text{I}}(x)=Ae^{ikx}+Be^{-ikx} \quad [1.198]$$

Zone II

For $E=-V_0$, the second term of the right member of equation [1.195] is zero. Hence:

$$\frac{d^2\Phi_{\text{II}}(x)}{dx^2}=0\Rightarrow\Phi_{\text{II}}(x)=C'x+D \quad [1.199]$$

C' and D are constant.

Since zone II starts at the connection point $x=a$, only the values of $x\geq a$ can be taken into account. Consequently the solution in zone II is written as:

$$\Phi_{\text{II}}(x)=C(x-a)+D \quad [1.200]$$

Zone III

Equation [1.196] admits the following solution:

$$\Phi_{III}(x) = Fe^{ik'x} + Ge^{-ik'x}$$

Since $E < 0$, let us consider $k' = iq$. This yields:

$$\Phi_{III}(x) = Fe^{-qx} + Ge^{qx}$$

As the wave function Φ_{III} must be square-summable (therefore bounded), the second term of the right member of the above equation is physically inconceivable. Hence $G = 0$. Moreover, as previously, let us change x into $x - 3a$. We obtain:

$$\Phi_{III}(x) = F\,e^{-q(x-3a)} \qquad [1.201]$$

(3) *Expression of the wave functions*

For $x < 0$, the wave function is zero, since the potential is infinite. Using [1.198], the connection conditions in $x = 0$ require $A + B = 0$. Hence: $A = -B$. The wave function $\Phi_I(x)$ is then written as:

$$\Phi_I(x) = A(e^{ikx} - e^{-ikx}) \Rightarrow \Phi_I(x) = 2A\sin kx \qquad [1.202]$$

According to [1.199], it can be noted that:

$$k^2 = 3q^2 \Rightarrow k = \sqrt{3}q \qquad [1.203]$$

Inserting [1.203] into [1.202], and considering $X = 2A$, we find:

$$\begin{cases} \Phi_I(x) = X\sin\left(\sqrt{3}qx\right) \\ \Phi_{II}(x) = C(x-a) + D \\ \Phi_{III}(x) = F\,e^{-q(x-3a)} \end{cases} \qquad [1.204]$$

(4) *Proof*

Let us express the boundary conditions in $x = a$, then in $x = 3a$. We have:

$$\begin{aligned}&\Phi_{\mathrm{I}}(a) = \Phi_{\mathrm{II}}(a);\ \Phi_{\mathrm{II}}(3a) = \Phi_{\mathrm{III}}(3a)\\ &\Phi'_{\mathrm{I}}(a) = \Phi'_{\mathrm{II}}(a);\ \Phi'_{\mathrm{II}}(3a) = \Phi'_{\mathrm{III}}(3a)\\ &\Phi'_{\mathrm{i}} = d\Phi/dx\end{aligned} \qquad [1.205]$$

Using [1.205] and [1.204], we have:

$$x = a : \begin{cases} X\sin\left(\sqrt{3}aq\right) = D \\ \sqrt{3}qX\cos\left(\sqrt{3}aq\right) = C \end{cases} ; x = 3a : \begin{cases} 2aC + D = F \\ C = -qF \end{cases} \qquad [1.206]$$

Using [1.206], the ratios D/C are written as:

$$\frac{D}{C} = \frac{1}{\sqrt{3}\,q}\tan\left(\sqrt{3}aq\right);\ \frac{D}{C} = -2a - \frac{1}{q} \qquad [1.207]$$

Using [1.207], we have:

$$\frac{1}{\sqrt{3}\,q}\tan\left(\sqrt{3}aq\right) = -2a - \frac{1}{q}$$

This leads to:

$$\tan\left(\sqrt{3}aq\right) = -2a\sqrt{3}\,q - \sqrt{3} \qquad [1.208]$$

Hence:

$$\tan Y = -2Y - \sqrt{3} \qquad [1.209]$$

with $Y = \sqrt{3}aq$.

(5) *Obtaining a bound state of energy*

The condition for obtaining a *bound state* of energy $E = -V_0$ is satisfied for all the solutions of equation [1.209]. The set of these solutions at a bound state of energy $E = -V_0$ is given by the points of intersection of the curves of equations:

$$y_1 = \tan Y \text{ and } y_2 = -2Y - \sqrt{3} \qquad [1.210]$$

(6) *Graphical solutions*

Some solutions of equations [1.210] are indicated in Figure 1.20.

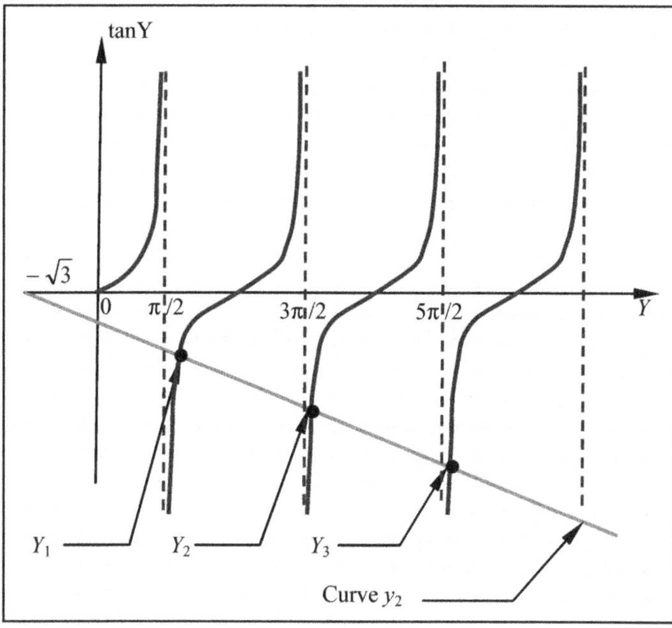

Figure 1.20. *Graphical solution to equation [1.210] providing the possible values of equation $Y = \sqrt{3}aq$, for which a bound state of energy $E = -V_0$ is reached. Only three solutions Y_1, Y_2 and Y_3 are indicated in the figure*

1.8.5. *Solution 5 – Particle confined in a rectangular potential*

(1) *Description*

From a classical point of view, the particle is confined in the rectangular well, since the potential is infinite outside of it. Moreover, its energy can have any value: its spectrum is continuous.

From the perspective of quantum mechanics, the wave function of the particle is zero outside of the dot. The evolution of the state of the particle inside the well is governed by the Schrödinger equation.

(2) *Differential equations*

Schrödinger's equation in zone I is written in two dimensions:

$$\left[-\frac{\hbar^2}{2m}\left(\frac{d^2}{dx}+\frac{d^2}{dy}\right)+V(x,y)\right]\Psi(x,y)=E\Psi(x,y) \qquad [1.211]$$

The potential being zero inside the rectangle, this equation becomes:

$$-\frac{\hbar^2}{2m}\left(\frac{\partial^2\Psi(x,y)}{\partial x}+\frac{\partial^2\Psi(x,y)}{\partial y}\right)=E\Psi(x,y) \qquad [1.212]$$

Let us use the variable separation method by writing the global wave function in the form: $\Psi\,(x, y) = \Phi\,(x) \times \psi\,(y)$. Equation [1.212] can be written as:

$$-\frac{\hbar^2}{2m}\left(\psi(y)\frac{d^2\Phi(x)}{dx}+\Phi(x)\frac{d^2\psi(y)}{dy}\right)=E\Phi(x)\psi(y) \qquad [1.213]$$

Dividing both sides of equation [1.212] by the functions $\Phi(x)$ and $\psi(y)$ and knowing that $E = E_x + E_y$= constant, we get:

$$-\frac{\hbar^2}{2m}\left(\frac{1}{\Phi(x)}\frac{d^2\Phi(x)}{dx}+\frac{1}{\psi(y)}\frac{d^2\psi(y)}{dy}\right)=E_x+E_y \qquad [1.214]$$

By identification, we find:

$$\begin{cases}-\dfrac{\hbar^2}{2m}\dfrac{1}{\Phi(x)}\dfrac{d^2\Phi(x)}{dx}=E_x\\[2ex]-\dfrac{\hbar^2}{2m}\dfrac{1}{\psi(y)}\dfrac{d^2\psi(y)}{dy}=E_y\end{cases} \qquad [1.215]$$

(3) *Solutions $\Phi_n\,(x)$ and $\psi_q(y)$*

Solutions $\Phi_n(x)$ and $\psi_q(y)$ to these equations are already known (see [1.129]). We find:

$$\begin{cases}\Phi_n(x)=\sqrt{\dfrac{2}{a}}\sin\left(\dfrac{n\pi x}{a}\right)\\[2ex]\psi_q(y)=\sqrt{\dfrac{2}{b}}\sin\left(\dfrac{q\pi y}{b}\right)\end{cases} \qquad [1.216]$$

In [1.216], the quantum numbers n and q are strictly positive. Since values $n = q = 0$ lead to zero energy, they are forbidden by the uncertainty principle.

(4) *Expression of the normed wave function, minimal values of n and q*

The normed wave function $\Psi_{nq}(x, y) = \Phi_n(x) \times \psi_q(y)$. Or, using [1.216]:

$$\Psi_{n,q}(x,y) = \frac{2}{\sqrt{ab}}\sin\left(\frac{n\pi x}{a}\right)\sin\left(\frac{q\pi y}{b}\right) \qquad [1.217]$$

(5) *Expression of the density of probability*

The density of the probability of finding the particle at point $M(x, y)$ is given by the square of the amplitude of probability $|\Psi_{nq}(x, y)|^2$. Hence:

$$D_{n,q}(x,y) = \frac{4}{ab}\sin^2\left(\frac{n\pi x}{a}\right)\sin^2\left(\frac{q\pi y}{b}\right) \qquad [1.218]$$

This density is maximal if the two functions $\sin^2(n\pi x/a)$ and $\sin^2(q\pi y/b)$ are simultaneously maximal. Therefore:

$n\pi x/a = (2k_x + 1)\,\pi/2$ $(1 \leq k_x \leq n - 1)$ and

$q\pi y/b = (2\,k_y + 1)\,\pi/2$ $(1 \leq k_y \leq q - 1)$

In the particular case of $k_x = k_y = 0$, we have:

$$\begin{cases} \sin^2\left(\dfrac{n\pi x}{a}\right) = 1 \Rightarrow \dfrac{n\pi x}{a} = \dfrac{\pi}{2} \\ \sin^2\left(\dfrac{q\pi y}{b}\right) = 1 \Rightarrow \dfrac{q\pi y}{b} = \dfrac{\pi}{2} \end{cases}$$

which finally leads to:

$$M(x_n, y_q) = \begin{cases} x_n = \dfrac{a}{2n} \\ y_m = \dfrac{b}{2q} \end{cases} \qquad [1.219]$$

(6) *Quantization of the total energy, case of square potential*

According to [1.62], we already know that for one dimension the total energy E is quantized. Designating the width of the well as l, we have:

$$E_n = \frac{\hbar^2\pi^2}{2ml^2}n^2 \qquad [1.220]$$

If, in this formula, we replace l with a or b as applicable, we get:

$$\begin{cases} E_n = \dfrac{\hbar^2\pi^2}{2ma^2}n^2 \\ E_q = \dfrac{\hbar^2\pi^2}{2mb^2}q^2 \end{cases} \Rightarrow E_{n,q} = \frac{\hbar^2\pi^2}{2m}\left(\frac{n^2}{a^2}+\frac{q^2}{b^2}\right) \qquad [1.221]$$

Result [1.221] shows that the spectrum of the particle is discrete, contrary to the predictions of classical mechanics, according to which the spectrum is continuous.

(7) *Square well, degeneracy of the energy levels*

For a square potential well ($a = b = l$), the energy E' is:

$$E'_{n,q} = \frac{\hbar^2\pi^2}{2ml^2}\left(n^2+q^2\right) \Rightarrow E'_{n,q} = E_0\left(n^2+q^2\right)$$

$$E_0 = \frac{\hbar^2\pi^2}{2ml^2} \qquad [1.222]$$

Expression [1.222] shows that there are various pairs of values (n, q) giving the same value of the energy $E'_{n,q}$. For this reason, the levels are said to be degenerate. To illustrate this, Table 1.2 summarizes the degree (or order) of degeneracy of these levels. The degree of degeneracy is the number of quantum states characterized by the same value of the energy of the system.

n	q	$E_{n,q}$	Degree of degeneracy ($g_{n,q}$)
1	1	E_0	$g_{1,1} = 1$: non-degenerate ground level
1	2	$5E_0$	$g_{1,2} = g_{2,1} = 2$: twice degenerate first excited level
2	1		
2	2	$8E_0$	$g_{2,2} = 1$: non-degenerate second excited level
1	3	$10E_0$	$g_{1,3} = g_{3,1} = 2$: twice degenerate third excited level
3	1		
2	3	$13E_0$	$g_{2,3} = g_{3,2} = 2$: twice degenerate fourth excited level
3	2		
3	3	$18E_0$	$g_{3,3} = 1$: non-degenerate fifth excited level

Table 1.2. *Degeneracy of the levels of energy of a particle confined in a square potential well*

The degeneracy of the levels of energy E' is due to the *symmetry* of potential for which $a = b = l$: the two axes Ox and Oy are therefore equivalent.

1.8.6. *Solution 6 – Square potential well: unbound states*

(1) *Behavior of the particle*

From a classical point of view, the particle passes over the well and carries on its uniform rectilinear motion. The particle has the same speed before reaching the well and after leaving it; let v_1 be this speed. Its speed v_2 above the well is constant:

$$v_1 = \sqrt{\frac{2E}{m}}\ ;\ v_2 = \sqrt{\frac{2(E+V_0)}{m}} \qquad [1.223]$$

From a quantum point of view, the state of the particle is described by a wave function. The particle has a non-zero probability of being reflected.

(2) *Schrödinger's equation, solutions $\Phi_I(x)$, $\Phi_{II}(x)$ and $\Phi_{III}(x)$*

– *Schrödinger's equation*

In zones I, II and III, Schrödinger's equation is written as, respectively:

$$\frac{d^2\Phi_{\text{I}}(x)}{dx^2} + \frac{2mE}{\hbar^2}\Phi_{\text{I}}(x) = 0$$

$$\frac{d^2\Phi_{\text{II}}(x)}{dx^2} + \frac{2m}{\hbar^2}(E + V_0)\Phi_{\text{II}}(x) = 0 \qquad [1.224]$$

$$\frac{d^2\Phi_{\text{III}}(x)}{dx^2} + \frac{2mE}{\hbar^2}\Phi_{\text{III}}(x) = 0$$

– *Solutions $\Phi_I(x)$, $\Phi_{II}(x)$ and $\Phi_{III}(x)$*

Since the backward wave is absent in zone III, then $G = 0$ in the expression of $\Phi_{\text{III}}(x)$. The solutions to the above equations are written as, respectively:

$$\begin{cases} \Phi_{\text{I}}(x) = Ae^{ikx} + Be^{-ikx} \\ \Phi_{\text{II}}(x) = De^{ipx} + De^{-ipx} \\ \Phi_{\text{III}}(x) = Fe^{ikx} \end{cases} \qquad [1.225]$$

In relations [1.225]:

$$k^2 = \frac{2mE}{\hbar^2};\ \rho^2 = \frac{2m(E+V_0)}{\hbar^2} \qquad [1.226]$$

To ease the calculation, the origin of coordinates undergoes translation:

$(x + a)$ for $x < -a$; $(x - a)$ for $x > + a$; $(x - a)$ for $-a \leq x \leq + a$

Solutions [1.227] are then written as:

$$\begin{cases} \Phi_{\text{I}}(x) = Ae^{ik(x+a)} + Be^{-ik(x+a)} \\ \Phi_{\text{II}}(x) = Ce^{i\rho(x-a)} + De^{-i\rho(x-a)} \\ \Phi_{\text{III}}(x) = Fe^{ik(x-a)} \end{cases} \qquad [1.227]$$

This writing clearly shows that at the well connection points in $x = -a$ and in $x = + a$, the exponential factors are equal to unity. This makes the calculations simpler.

(3) *Probability of transmission*

Let us express the boundary conditions in $x = -a$ and then in $x = +a$:

$$\Phi_{\text{I}}(-a) = \Phi_{\text{II}}(-a);\ \Phi_{\text{II}}(a) = \Phi_{\text{III}}(a)$$

$$\Phi'_{\text{I}}(-a) = \Phi'_{\text{II}}(-a);\ \Phi'_{\text{II}}(a) = \Phi'_{\text{III}}(a) \qquad [1.228]$$

$$\Phi'_{\text{i}} = d\Phi/dx$$

Using [1.227] and [1.228], we find:

$$x = -a: \begin{cases} A + B = Ce^{-2i\rho a} + De^{2i\rho a} \\ ikA - ikB = i\rho Ce^{-2i\rho a} - i\rho De^{2i\rho a} \end{cases} \qquad [1.229]$$

$$x = a: \begin{cases} C + D = F \\ i\rho C - i\rho D = ikF \end{cases} \qquad [1.230]$$

The transmission coefficient is defined by the relation (knowing that $k_{\text{III}} = k_{\text{I}} = k$):

$$T = \left|\frac{F}{A}\right|^2 \times \frac{k_{III}}{k_I} \Rightarrow T = \left|\frac{F}{A}\right|^2 \qquad [1.231]$$

Using [1.230], we express C and D as functions of F, and A as a function of C and D. We then have:

$$\begin{cases} 2C = \left(\dfrac{\rho+k}{\rho}\right)F \\ 2D = \left(\dfrac{\rho-k}{\rho}\right)F \end{cases} ;2A = C\left(\frac{\rho+k}{k}\right)e^{-2i\rho a} - D\left(\frac{\rho-k}{k}\right)e^{2i\rho a} \qquad [1.232]$$

Using [1.232], we have:

$$2A = \frac{F}{2}\left(\frac{\rho+k}{\rho}\right)\left(\frac{\rho+k}{k}\right)e^{-2i\rho a} - \frac{F}{2}\left(\frac{\rho-k}{\rho}\right)\left(\frac{\rho-k}{k}\right)e^{2i\rho a}$$

which is:

$$\frac{A}{F} = \frac{1}{4\rho k}\left[(\rho+k)^2 e^{-2i\rho a} - (\rho-k)^2 e^{2i\rho a}\right]$$

Expanding the expression between brackets and then simplifying, we have:

$$\frac{A}{F} = \cos 2\rho a - \frac{i}{2}\frac{\left(\rho^2+k^2\right)}{\rho k}\sin 2\rho a \qquad [1.233]$$

Knowing that $\cos^2 x + \sin^2 x = 1$, we get:

$$\left|\frac{A}{F}\right|^2 = 1 - \sin^2 2\rho a + \frac{1}{4}\frac{\left(\rho^2+k^2\right)^2}{\rho^2 k^2}\sin^2 2\rho a$$

Factorizing the second member by $\sin^2 2\rho a$, after arrangement we obtain:

$$\left|\frac{A}{F}\right|^2 = \frac{4\rho^2 k^2 + \left[\left(\rho^2+k^2\right)^2 - 4\rho^2 k^2\right]\sin^2 2\rho a}{4\rho^2 k^2}$$

or after expansion and simplification of the terms between brackets:

$$\left|\frac{A}{F}\right|^2 = \frac{4\rho^2 k^2 + \left(\rho^2 - k^2\right)^2 \sin^2 2\rho a}{4\rho^2 k^2} \qquad [1.234]$$

The inverse of relation [1.234] gives the expression of the barrier transmission. We finally have:

$$T = \frac{4\rho^2 k^2}{4\rho^2 k^2 + \left(\rho^2 - k^2\right)^2 \sin^2 2\rho a} \qquad [1.235]$$

Using the expressions of k^2 and ρ^2 according to [1.226], we express the transmission coefficient T as a function of E and V_0. Hence:

$$T = \frac{4(E+V_0)}{4(E+V_0) + V_0^2 \sin^2\left[\frac{\sqrt{8m(E+V_0)}}{\hbar} a\right]} \qquad [1.236]$$

(4) *Definition, expression*

Taking [1.230] into account, the reflection probability $R = |B/A|$ is written as:

$$R = \left|\frac{B}{F}\right|^2 \times \left|\frac{F}{A}\right|^2 \qquad [1.237]$$

or as a function of the barrier transmission:

$$R = \left|\frac{B}{F}\right|^2 T \qquad [1.238]$$

It is then sufficient to express the ratio B/F. Using [1.237], we express B as a function of C and D. We obtain:

$$2B = C\left(\frac{k-\rho}{k}\right)e^{-2i\rho a} + D\left(\frac{k+\rho}{k}\right)e^{2i\rho a} \qquad [1.239]$$

Using the first relations [1.232], relation [1.239] is written as:

$$2B = \frac{F}{2}\left(\frac{k-\rho}{k}\right)\left(\frac{k+\rho}{\rho}\right)e^{-2i\rho a} - \frac{F}{2}\left(\frac{\rho+k}{k}\right)\left(\frac{k-\rho}{\rho}\right)e^{2i\rho a}$$

which is:

$$\frac{B}{F} = \frac{k^2-\rho^2}{4\rho k}\left[e^{-2i\rho a} - e^{2i\rho a}\right]$$

Expanding the term between brackets, we have:

$$\frac{B}{F}=-\frac{i}{2}\frac{k^2-\rho^2}{\rho k}\sin 2\rho a$$

Using the last equality, we have:

$$\left|\frac{B}{F}\right|^2=\frac{\left(k^2-\rho^2\right)^2}{4\rho^2k^2}\sin^2 2\rho a \qquad [1.240]$$

Inserting [1.240] and [1.236] into [1.238], after simplification we get:

$$R=\frac{\left(k^2-\rho^2\right)^2\sin^2 2\rho a}{4\rho^2k^2+\left(\rho^2-k^2\right)^2\sin^2 2\rho a} \qquad [1.241]$$

(5) *Predictions*

We determine the sum $T+R$. Using [1.237] and [1.242], we obtain:

$$T+R=\frac{4\rho^2k^2}{4\rho^2k^2+\left(\rho^2-k^2\right)^2\sin^2 2\rho a}+\frac{\left(k^2-\rho^2\right)^2\sin^2 2\rho a}{4\rho^2k^2+\left(\rho^2-k^2\right)^2\sin^2 2\rho a}$$

or, after arrangement and expansion:

$$T+R=\frac{4\rho^2k^2+\left(k^4+\rho^4-2k^2\rho^2\right)\sin^2 2\rho a}{4\rho^2k^2+\left(\rho^2-k^2\right)^2\sin^2 2\rho a} \qquad [1.242]$$

Which gives after simplification: $T+R=1$.

CONCLUSION.– From a quantum point of view, the particle is either reflected or transmitted. The predictions of classical mechanics are quite different: the particle is transmitted without being able to turn back.

(6) *Spectrum of the particle*

The energy spectrum of the particle is continuous. This is due to the fact that the states of energy $E>0$ are unbound states.

1.8.7. *Solution 7 – Square potential well: bound states*

(1) *Wave functions*

Let us consider:

$$\rho^2 = -\frac{2mE}{\hbar^2};\ k^2 = \frac{2m(E+V_0)}{\hbar^2} \qquad [1.243]$$

Schrödinger's equations in zones I, II and III are given by the previous relations [1.224].

Solutions $\Phi_I(x)$, $\Phi_{II}(x)$ and $\Phi_{III}(x)$ to the above equations are the following (to facilitate the calculation, x changes into $x + a$ for $x \leq -a$ or $x - a$ for $x \geq a$ for the solutions in zones I and III):

$$\begin{cases} \Phi_I(x) = Ae^{ik'(x+a)} + Be^{-ik'(x+a)} \\ \Phi_{II}(x) = Ce^{ikx} + De^{-ik} \\ \Phi_{III}(x) = Ae^{ik'(x-a)} + Be^{-ik'(x-a)} \end{cases} \qquad [1.244]$$

Since $E < V_0$, we consider $k' = i\rho$. We obtain:

$$\begin{cases} \Phi_I(x) = Ae^{-\rho(x+a)} + Be^{\rho(x+a)} \\ \Phi_{II}(x) = Ce^{ikx} + De^{-ik} \\ \Phi_{III}(x) = Fe^{-\rho(x-a)} + Ge^{\rho(x-a)} \end{cases} \qquad [1.245]$$

The wave function being bounded, then for $x \leq -a$, the function $Ae^{-\rho(x+a)}$ is divergent. Moreover, for $x \geq a$, the wave function $Ge^{\rho(x-a)}$ is also divergent. Consequently, we must simultaneously have $A = 0$ and $G = 0$. In summary, solutions [1.245] are written as:

$$\begin{cases} \Phi_I(x) = Be^{\rho(x+a)} \\ \Phi_{II}(x) = Ce^{ikx} + De^{-ik} \\ \Phi_{III}(x) = Fe^{-\rho(x-a)} \end{cases} \qquad [1.246]$$

(2) *Proof, energy quantization*

– *Proof*

Let us express the connection conditions of the wave function in $x = -a$ and then in $x = +a$, which are:

$$\Phi_{\rm I}(-a) = \Phi_{\rm II}(-a);\ \Phi_{\rm II}(a) = \Phi_{\rm III}(a)$$

$$\Phi'_{\rm I}(-a) = \Phi'_{\rm II}(-a);\ \Phi'_{\rm II}(a) = \Phi'_{\rm III}(a) \qquad [1.247]$$

$$\Phi'_{\rm i} = d\Phi/dx$$

Using [1.247] and [1.246], we find:

$$x=-a:\begin{cases} B = Ce^{-ika}+De^{ika} \\ \rho B = ikCe^{-ika}-ikDe^{ika} \end{cases} \qquad [1.248]$$

$$x=a:\begin{cases} Ce^{ika}+De^{-ika} = F \\ ikCe^{ika}-ikDe^{-ika}=-\rho F \end{cases} \qquad [1.249]$$

Let us express the coefficients C and D as a function of B. Considering the first system of equations [1.249], we multiply the two equations by e^{ika}. Then we divide the second equation by ik and multiply the obtained result by e^{ika}. We obtain:

$$\begin{cases} e^{ika}B = C + De^{2ika} \\ e^{ika}\dfrac{\rho B}{ik} = C - De^{2ika} \end{cases} \qquad [1.250]$$

Similarly, we multiply the first equation in [1.249] by e^{-ika}. We then divide the second equation by $-ik$. Multiplying the obtained result by e^{-ika}, we find:

$$\begin{cases} e^{-ika}B = Ce^{-2ika}+D \\ -e^{-ika}\dfrac{\rho B}{ik} = -Ce^{-2ika}+D \end{cases} \qquad [1.251]$$

Using [1.250] and [1.251], we have:

$$\begin{cases} 2C = \left(1+\dfrac{\rho}{ik}\right)e^{ika}B = \left(\dfrac{\rho+ik}{ik}\right)e^{ika}B \\ 2D = \left(1-\dfrac{\rho}{ik}\right)e^{-ika}B = -\left(\dfrac{\rho-ik}{ik}\right)e^{-ika}B \end{cases} \qquad [1.252]$$

$$\begin{cases} C = \dfrac{k - i\rho}{2k} e^{ika} B \\ D = \dfrac{k + i\rho}{2k} e^{-ika} B \end{cases} \qquad [1.253]$$

We then express F as a function of C and D. For this purpose, we use the second system of equations [1.249] and then proceed as previously. We have:

$$\begin{cases} C + De^{-2ika} = Fe^{-ika} \\ C - De^{-2ika} = -\dfrac{\rho}{ik} Fe^{-ika} \end{cases} \Rightarrow \begin{cases} Ce^{2ika} + D = Fe^{ika} \\ -Ce^{2ika} + D = \dfrac{\rho}{ik} Fe^{ika} \end{cases}$$

From these equations, we deduce:

$$\begin{cases} 2C = \left(1 - \dfrac{\rho}{ik}\right) e^{-ika} F = -\left(\dfrac{\rho - ik}{ik}\right) e^{-ika} F \\ 2D = \left(1 + \dfrac{\rho}{ik}\right) e^{ika} F = \left(\dfrac{\rho + ik}{ik}\right) e^{ika} F \end{cases} \qquad [1.254]$$

Using [1.254], we express F as a function of C and then as a function of D. Hence:

$$\begin{cases} F = -2\left(\dfrac{ik}{\rho - ik}\right) e^{ika} C \\ F = 2\left(\dfrac{ik}{\rho + ik}\right) e^{-ika} D \end{cases} \qquad [1.255]$$

It is easy to express F as a function of B using [1.253]. This leads to:

$$\begin{cases} F = -\left(\dfrac{\rho + ik}{\rho - ik}\right) e^{2ika} B \\ F = -\left(\dfrac{\rho - ik}{\rho + ik}\right) e^{-2ika} B \end{cases} \qquad [1.256]$$

Equalizing the two relations [1.256], after arrangement we find:

$$\left(\frac{\rho - ik}{\rho + ik}\right)^2 = e^{4ika} \qquad [1.257]$$

– *Quantization of energy*

Let us recall the relations [1.243]:

$$\rho^2 = -\frac{2mE}{\hbar^2}\,;\; k^2 = \frac{2m(E+V_0)}{\hbar^2}$$

These relations prove that ρ and k depend on the energy E. Consequently, equation [1.257] can only be satisfied for certain values of E: the energy is therefore quantized (the possible values are discrete). Hence, requiring the wave function to be square-summable in zones I and III drives a *quantization* of the energy of the particle.

(3) *Graphical solution, parity of the wave functions*

Two cases are possible:

$$\left(\frac{\rho - ik}{\rho + ik}\right) = -e^{2ika};\quad \left(\frac{\rho - ik}{\rho + ik}\right) = e^{2ika} \qquad [1.258]$$

The first equation of [1.258] gives:

$$\rho - ik = -\,\rho e^{2ika} - ike^{2ika}$$

This yields:

$$\rho\left(1 + e^{2ika}\right) = ik\left(1 - e^{2ika}\right)$$

which is:

$$\frac{\rho}{k} = i\frac{\left(1 - e^{2ika}\right)}{\left(1 - e^{2ika}\right)} = i\frac{\left(e^{-ika} - e^{ika}\right)}{\left(e^{-ika} + e^{ika}\right)}$$

Expanding the terms between brackets in the second member of the above expression, we have:

$$\frac{\rho}{k} = \tan(ka) \qquad [1.259]$$

where $\tan(ka) > 0$.

The solution to equation [1.259] is not convenient due to the ratio ρ/k, since the parameters ρ and k both depend on the energy E.

To find a more convenient equation to solve, we consider relations [1.243]. We then note that:

$$k^2 = \frac{2mE}{\hbar^2} + \frac{2mV_0}{\hbar^2} = -\rho^2 + \frac{2mV_0}{\hbar^2}$$

which is:

$$k^2 + \rho^2 = \frac{2mV_0}{\hbar^2} \qquad [1.260]$$

For $E = 0$, $\rho = 0$ and $k = k_0$. Equation [1.260] can then be written in the following form:

$$k_0^2 = \frac{2mV_0}{\hbar^2} \Rightarrow k_0 = \sqrt{k^2 + \rho^2} \qquad [1.261]$$

The constant k_0, independent of the energy E, makes it possible to set a simple equation, the graphic solution of which is easy. For this purpose, we also note that:

$$\frac{1}{\cos^2 ka} = 1 + \tan^2 ka$$

Considering relation [1.259], we thus obtain:

$$\frac{1}{\cos^2 ka} = 1 + \frac{\rho^2}{k^2} = \frac{k^2 + \rho^2}{k^2} \Rightarrow \frac{1}{\cos^2 ka} = \frac{k_0^2}{k^2} = \left(\frac{k_0}{k}\right)^2$$

Using the last equality, we finally get:

$$\begin{cases} |\cos ka| = \dfrac{k}{k_0} \\ \tan(ka) > 0 \end{cases} \qquad [1.262]$$

The system of equations [1.262] is equivalent to equation [1.259]. The function coska being even, equation [1.262] is verified provided that tan $(ka) > 0$. The discrete values of energy E are then given by the intersection of a sine function y_1 and the line y_2 given by the expressions:

$$\begin{cases} y_1 = |\cos ka| \\ y_2 = \dfrac{k}{k_0} \\ \tan(ka) > 0 \end{cases} \qquad [1.263]$$

We now study the second scenario, considering [1.258]:

$$\left(\frac{\rho - ik}{\rho + ik}\right) = e^{2ika} \qquad [1.264]$$

Expanding this equation as previously, we get:

$$\frac{k}{\rho} = -\tan(ka) \qquad [1.265]$$

with tan $(ka) < 0$.

We consider the trigonometric transformation:

$$\cos^2 ka = \frac{1}{1 + \tan^2 ka} = 1 - \sin^2 ka$$

Equation [1.264] can thus be written as:

$$\sin^2 ka = 1 - \frac{\rho^2}{\rho^2 + k^2}$$

After arrangement, the above relation becomes:

$$\sin^2 ka = \frac{k^2}{\rho^2 + k^2}$$

Taking [1.261] into account, we have:

$$\sin^2 ka = \left(\frac{k}{k_0}\right)^2 \Rightarrow \begin{cases} |\sin ka| = \dfrac{k}{k_0} \\ \tan(ka) < 0 \end{cases} \qquad [1.266]$$

The function sin*ka* is odd, hence equation [1.266] is satisfied with the additional condition tan $(ka) < 0$. The discrete values of energy E are given by the intersection of a sine function y_1 and the line y_2 given by the following expressions:

$$\begin{cases} y_1 = |\sin ka| \\ y_2 = \dfrac{k}{k_0} \\ \tan(ka) < 0 \end{cases} \qquad [1.267]$$

(4) Let us now solve equation [1.262] graphically using the equivalent equations [1.266] and [1.267]. We obtain the curves represented in Figure 1.21.

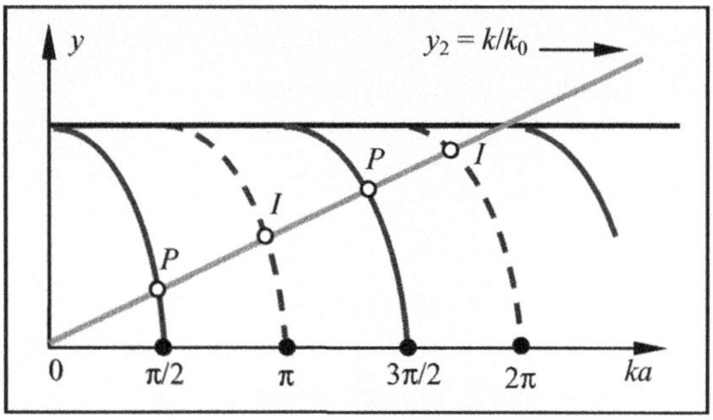

Figure 1.21. *Graphic solution of equation [1.257] giving the discrete values of energy of the bound states of a particle confined in a square potential well. The values at the points of intersection P (thick line curve) correspond to the solutions to equation [1.263], and those at the points of intersection I (dotted line curve) correspond to the solutions to equation [1.267]*

For $0 < ka < \pi/2$, $\pi < ka < 3\pi/2$, etc., we have tan $ka > 0$. This satisfies the condition imposed for solution [1.263] and the corresponding curves are represented in thick line in Figure 1.21.

Similarly, for $\pi/2 < ka < \pi$, $3\pi/2 < ka < 2\pi$, etc., we have tan $ka < 0$. This meets the condition imposed to solution [1.267] and the corresponding curves are represented in dotted lines in Figure 1.21. Only four bound states are marked in this figure. Two of them are associated with points P and the other two with points I.

– Parity of the wave function

To prove that the associated wave functions are either even or odd, we express the ratios C/D and F/B using the relations [1.253] and [1.256]; then we take [1.258] into account. Hence:

$$\begin{cases} \dfrac{C}{D} = \left(\dfrac{k - i\rho}{k + i\rho}\right) e^{2ika} \\ \dfrac{F}{B} = -\left(\dfrac{\rho + ik}{\rho - ik}\right) e^{2ika} \end{cases} \Rightarrow \begin{cases} \left(\dfrac{\rho - ik}{\rho + ik}\right) = -e^{2ika} \\ \left(\dfrac{\rho - ik}{\rho + ik}\right) = e^{2ika} \end{cases} \quad [1.268]$$

– *Even wave functions*

Using the first relations of systems [1.268], we get:

$$\begin{cases} \dfrac{C}{D} = -\left(\dfrac{k - i\rho}{k + i\rho}\right) \times \left(\dfrac{\rho - ik}{\rho + ik}\right) = 1 \\ \dfrac{F}{B} = \left(\dfrac{\rho + ik}{\rho - ik}\right) \times \left(\dfrac{\rho - ik}{\rho + ik}\right) = 1 \end{cases} \quad [1.269]$$

NOTE.– We have multiplied both the numerator and the denominator of the first equation [1.269] by the imaginary number i; this effectively leads to $C/D = 1$:

$$\frac{C}{D} = \left(\frac{\rho + ik}{\rho - ik}\right) \times \left(\frac{\rho - ik}{\rho + ik}\right) = 1$$

Knowing that $C = D$ and $F = B$ according to [1.269], the wave functions [1.248] are:

$$\begin{cases} \Phi_{\text{I}}(x) = Be^{\rho(x+a)} \\ \Phi_{\text{II}}(x) = C\left(e^{ikx} + e^{-ik}\right) \\ \Phi_{\text{III}}(x) = Be^{-\rho(x-a)} \end{cases} \quad [1.270]$$

A wave function is even if $\Psi(-x) = \Psi(x)$. Given [1.270], we have:

$$\begin{cases} \Phi_{\text{I}}(-x) = Be^{\rho(-x+a)} = Be^{-\rho(x-a)} = \Phi_{\text{III}}(x) \\ \Phi_{\text{II}}(-x) = C\left(e^{ikx} + e^{-ikx}\right) = \Phi_{\text{II}}(x) \\ \Phi_{\text{III}}(-x) = Be^{-\rho(-x-a)} = Be^{\rho(x+a)} = \Phi_{\text{I}}(x) \end{cases} \quad [1.271]$$

Using [1.271] it can be verified that $\Phi_i(-x) = \Phi_i(x)$: the bound states associated with the energies corresponding to the solutions to equation [1.263] are therefore even (in other words, the *wave functions* are *symmetrical*). Figure 1.21 shows the two even bound states associated with points P (therefore this point marks the even solution).

– *Odd wave functions*

Using the two relations of systems [1.268], and proceeding as previously, we get:

$$\begin{cases} \dfrac{C}{D} = \left(\dfrac{k - i\rho}{k + i\rho}\right) \times \left(\dfrac{\rho - ik}{\rho + ik}\right) = -1 \\ \dfrac{F}{B} = -\left(\dfrac{\rho + ik}{\rho - ik}\right) \times \left(\dfrac{\rho - ik}{\rho + ik}\right) = -1 \end{cases} \qquad [1.272]$$

Using these results, the wave functions [1.270] are written as follows:

$$\begin{cases} \Phi_{\rm I}(x) = -B\, e^{\rho(x+a)} \\ \Phi_{\rm II}(x) = -C\left(e^{ikx} + e^{-ikx}\right) \\ \Phi_{\rm III}(x) = -B\, e^{-\rho(x-a)} \end{cases} \qquad [1.273]$$

A wave function is considered odd if $\Psi(-x) = -\Psi(x)$. Considering [1.273], it can be verified that $\Phi_i(-x) = -\Phi_i(x)$: the bound states associated with energies corresponding to solutions to equation [1.267] are therefore *odd* (which corresponds to *antisymmetric wave functions*). As Figure 1.21 shows, the two odd bound states are associated with points *I* (therefore this point marks the odd solution).

1.8.8. *Solution 8 – Infinitely deep rectangular potential well*

(1) *Determination of the wave function*

The particle is confined to zone II where the potential is zero. It is in this zone that the Schrödinger equation must be written (it is already known that the wave function is zero in the other zones, where the potential is infinite). Then we obtain:

$$\frac{d^2\Phi_{\rm II}(x)}{dx^2} + \frac{2mE}{\hbar^2}\Phi_{\rm II}(x) = 0 \qquad [1.274]$$

The solution to this equation has the form:

$$\Phi_{\rm II}(x) = Ae^{ikx} + Be^{-ikx} \qquad [1.275]$$

(2) *Condition imposed on k, quantization of the energy*

– *Condition imposed on k*

Let us express the connection conditions of the wave function for $x = -a/2$ and then for $x = +a/2$: $\Phi_I(-a/2) = \Phi_{II}(-a/2) = 0$; $\Phi_{II}(a/2) = \Phi_{III}(a/2) = 0$. Using [1.275] this leads to:

$$\begin{cases} x=-\dfrac{a}{2}: Ae^{-ika/2}+Be^{ika/2}=0 \\ x=\dfrac{a}{2} \ : Ae^{ika/2}+Be^{-ika/2}=0 \end{cases}$$

These equations are transformed as follows:

$$\begin{cases} A\left(\cos\dfrac{ka}{2}-i\sin\dfrac{ka}{2}\right)+B\left(\cos\dfrac{ka}{2}+i\sin\dfrac{ka}{2}\right)=0 \\ A\left(\cos\dfrac{ka}{2}+i\sin\dfrac{ka}{2}\right)+B\left(\cos\dfrac{ka}{2}-i\sin\dfrac{ka}{2}\right)=0 \end{cases}$$

After factorization we have:

$$\begin{cases} (A+B)\cos\dfrac{ka}{2}+i(B-A)\sin\dfrac{ka}{2}=0 \\ (A+B)\cos\dfrac{ka}{2}+i(A-B)\sin\dfrac{ka}{2}=0 \end{cases} \qquad [1.276]$$

This system of equations is solved if: $(A + B) = 0$ and $(A - B) = 0$. This gives:

$$\begin{cases} A=-B \Rightarrow \sin\dfrac{ka}{2}=0 \\ A=B \Rightarrow \cos\dfrac{ka}{2}=0 \end{cases} \qquad [1.277]$$

The system [1.277] admits the following solutions:

$$\begin{cases} \sin\dfrac{ka}{2}=0 \Rightarrow \dfrac{ka}{2}=\dfrac{n\pi}{2} \Rightarrow k_n=\dfrac{n\pi}{a}, \quad n \text{ even} \\ \cos\dfrac{ka}{2}=0 \Rightarrow \dfrac{ka}{2}=\dfrac{n\pi}{2} \Rightarrow k_n=\dfrac{n\pi}{a}, \quad n \text{ odd} \end{cases} \qquad [1.278]$$

Finally, the condition imposed on k is written as:

$$k_n = \frac{n\pi}{a} \quad [1.279]$$

where n is a strictly positive integer.

– *Quantization of energy*

Knowing that $k^2 = 2mE/\hbar^2$, relation [1.279] gives:

$$k_n^2 = \frac{n^2\pi^2}{a^2} = \frac{2mE_n}{\hbar^2} \Rightarrow E_n = \frac{\hbar^2\pi^2}{2ma^2}n^2 \text{, } n \geq 1 \quad [1.280]$$

(3) *Expressions of wave functions, normalization*

– *Expressions of wave functions*

The even and odd wave functions describing the bound states of the particle are deduced from solutions [1.277]. Let us study the two possible cases.

First case: $A = B$

In this case, expression [1.275] gives:

$$\Phi_{\text{I}}(x) = A\left(e^{ik_n x} + e^{-ik_n x}\right) \Rightarrow \Phi_n(x) = 2A\cos k_n x$$

Or taking condition [1.279] into account:

$$\Phi_n(x) = 2A\cos\left(\frac{n\pi}{a}x\right) \quad [1.281]$$

In this expression, the integer n is odd according to [1.278]. The wave function [1.281] is *even* (therefore *symmetric*) since $\Phi_n(-x) = \Phi_n(x)$.

Second case: $A = -B$

In this case, expression [1.275] is written as:

$$\Phi_{\text{II}}(x) = A\left(e^{ik_n x} - e^{-ik_n x}\right) \Rightarrow \Phi_n(x) = 2iA\cos k_n x$$

or taking condition [1.279] into account:

$$\Phi_n(x) = 2iA\sin\left(\frac{n\pi}{a}x\right) \qquad [1.282]$$

In expression [1.278], the quantum number n is even and the wave function is *odd* (therefore *antisymmetric*). It can be verified that $\Phi_n(-x) = -\Phi_n(x)$. In summary, let us consider $C = 2A$ and $D = 2iA$. The wave function $\Phi_{\text{II}}(x)$ satisfies the following conditions:

$$\Phi_{\text{II}}(x) = \begin{cases} \Phi_n(x) = C\cos\left(\dfrac{n\pi}{a}x\right) \\ \Phi_n(x) = D\sin\left(\dfrac{n\pi}{a}x\right) \end{cases} \qquad [1.283]$$

It is worth recalling that the integer n is even for the first function and odd for the second function. Outside of the well, $\Phi(x) = 0$. Operating a translation of the origin of coordinates such that $x' = x - a/2$, expressions [1.283] are written as:

$$\Phi_{\text{II}}(x) = \begin{cases} \Phi_n(x) = C\cos\left(\dfrac{n\pi}{a}x - \dfrac{n\pi}{2}\right) \\ \Phi_n(x) = D\sin\left(\dfrac{n\pi}{a}x - \dfrac{n\pi}{2}\right) \end{cases} \qquad [1.284]$$

Knowing that $\cos(n\pi/2) = 0$ (n is odd) and $\sin(n\pi/2) = 0$ (n is even), we have:

$$\Phi_{\text{II}}(x) = \begin{cases} \Phi_n(x) = \pm C\sin\left(\dfrac{n\pi}{a}x\right) \\ \Phi_n(x) = \pm D\sin\left(\dfrac{n\pi}{a}x\right) \end{cases} \qquad [1.285]$$

with $n = 1, 2, 3, \ldots$.

– *Normed wave functions*

Using wave functions [1.283] and applying the normalization condition, and knowing that the particle is confined to the well, we have:

$$|C|^2 \int_{-a/2}^{a/2} \cos^2\left(\frac{n\pi}{a}x\right)dx = 2|C|^2 \int_0^{a/2} \cos^2\left(\frac{n\pi}{a}x\right)dx = 1$$

$$|D|^2 \int_{-a/2}^{a/2} \sin^2\left(\frac{n\pi}{a}x\right)dx = 2|D|^2 \int_0^{a/2} \sin^2\left(\frac{n\pi}{a}x\right)dx = 1$$

After integration this leads to:

$$\begin{cases} 2|C|^2 \times \frac{a}{4} = 1 \Rightarrow C = \sqrt{\frac{2}{a}} \\ 2|D|^2 \times \frac{a}{4} = 1 \Rightarrow D = \sqrt{\frac{2}{a}} \end{cases} \qquad [1.286]$$

Using results [1.286], the normed wave functions are written as follows:

$$\Phi_{II}(x) = \begin{cases} \Phi_n(x) = \sqrt{\frac{2}{a}} \cos\left(\frac{n\pi}{a}x\right) \\ \Phi_n(x) = \sqrt{\frac{2}{a}} \sin\left(\frac{n\pi}{a}x\right) \end{cases} \qquad [1.287]$$

Let us note that in [1.287], C and D have been chosen real and positive. Complex numbers can be chosen, which involves multiplying the factors in equations [1.288] by i; this does not change the physical predictions (*two proportional wave functions represent the same physical state*).

Moreover, physical predictions involve the *density of probability*; this eliminates the number i. As a general rule, it is always possible to choose real and positive normalization constants.

(4) *Graphical representation*

Let us express the wave functions and the densities of probability for the ground level ($n = 1$) and for the first three excited levels ($n = 2$; $n = 3$; $n = 4$) of the particle. These are summarized in Table 1.3 (where "S" stands for the *symmetric wave function* and "AS" for the *antisymmetric wave function*).

The graphical representation of the wave functions [1.287] and of the corresponding probability densities poses no difficulty (see Figure 1.3).

N	Wave function $\Phi_n(x)$	Density of probability $\lvert\Phi_n(x)\rvert^2$	Symmetry
1	$\sqrt{\frac{2}{a}}\cos\frac{\pi}{a}x$	$\frac{2}{a}\cos^2\frac{\pi}{a}x$	S
2	$\sqrt{\frac{2}{a}}\sin\frac{2\pi}{a}x$	$\frac{2}{a}\sin^2\frac{2\pi}{a}x$	AS
3	$\sqrt{\frac{2}{a}}\cos\frac{3\pi}{a}x$	$\frac{2}{a}\cos^2\frac{3\pi}{a}x$	S
4	$\sqrt{\frac{2}{a}}\sin\frac{4\pi}{a}x$	$\frac{2}{a}\sin^2\frac{4\pi}{a}x$	AS

Table 1.3. *Parity of the wave function describing the bound states of a particle confined in an infinitely deep square potential well*

1.8.9. *Solution 9 – Metal assimilated to a potential well, cold emission*

Let us consider an electron of energy W at point M of abscissa x_1 (*barrier thickness at point M*) tunneling out of the metal.

(1) *Determination of the expression of the potential energy*

As shown in Figure 1.17, the potential energy $V(x)$ varies linearly with x between B and D for $x > 0$. Hence: $V(x) = ax + b$; slope a is negative, as the function $V(x)$ is decreasing.

The potential energy $V(x)$ is the sum of the potential energy K outside the metal and the potential energy $E_p(x)$ is due to the interaction between the electron and the electric field of intensity E. By definition:

$$E_p(x) = qEx = -eEx \Rightarrow V(x) = -eEx + K \qquad [1.288]$$

In summary, the potential energy function $V(x)$ varies as follows:

$$V(x) = \begin{cases} 0, & x < 0 \\ -eEx + K, & x > 0 \end{cases} \qquad [1.289]$$

(2) *Expression of* x_1

The thickness x_1 of the barrier in M is determined by the condition $V(x_1) = W$ (Figure 1.22). Or, using [1.290]:

$$-eEx_1 + K = W_x \Rightarrow x_1 = \frac{K - W_x}{eE} \quad [1.290]$$

(3) *Proof, numerical applications*

The *transmission probability* T of the barrier is given by the expression:

$$T = \exp -2\int_0^{x_1} \sqrt{\frac{2m[V(x) - W_x]}{\hbar^2}}dx \quad [1.291]$$

Using [1.289], we obtain for $x > 0$ and knowing that $\phi = K - W$:

$$T = \exp -2\int_0^{x_1} \sqrt{\frac{2m[\phi - eEx]}{\hbar^2}}\, dx \quad [1.292]$$

In the exponential of [1.293], we insert:

$$I = 2\sqrt{2m}\int_0^{x_1} \sqrt{\frac{[\phi - eEx]}{\hbar^2}}\, dx \quad [1.293]$$

Integration leads to:

$$I = -\frac{2\sqrt{2m}}{3eE}\left[(\phi - eEx_1)^{3/2} - \phi^{3/2}\right] \quad [1.294]$$

Since $\phi = K - W$, according to [1.290] we have $x_1 = \phi/eE \Rightarrow \phi = eEx_1$. Result [1.294] is then written as:

$$I = \frac{2\sqrt{2m}\,\phi^{3/2}}{3eE} \Rightarrow T = \exp -\frac{4\sqrt{2m\phi^3}}{3e\hbar} \times \frac{1}{E} \quad [1.295]$$

Finally, this leads to:

$$T = \exp - E_0/E \quad [1.296]$$

with:

$$E_0 = \frac{4\sqrt{2m\phi^3}}{3e\hbar}$$

N.A.– $E = 4 \times 109$ V · m^{-1}; $E_0 = 7.66 \times 10^{10}$ V · m^{-1}; $T \approx 5 \times 10^{-9}$.

CONCLUSION.– The probability of cold emission of electrons from a metal is very low. On the other hand, this probability is measurable, since the number of electrons is very large.

1.8.10. *Solution 10 – Ground state energy of the harmonic oscillator*

(1) *Expression of the elastic potential energy*

The oscillator under consideration is subjected at any instant to an opposing spring force $F = -kx$. This force derives from the potential energy (it can be verified that $\vec{\nabla} \wedge \vec{F} = \vec{0}$):

$$V(x) = -\int F dx = k \int x dx \Rightarrow V(x) = \frac{1}{2} kx^2 \qquad [1.297]$$

The origin of potential energies has been chosen at point O, origin of coordinates, and the integration constant in [1.298] is therefore zero.

(2) *Proof*

By definition, the *classical harmonic oscillator* is a conservative system if its mechanical energy E is constant. By definition, $E = E_c + V(x)$. Hence:

$$E = \frac{1}{2} m \left(\frac{dx}{dt} \right)^2 + \frac{1}{2} kx^2$$

Differentiating this expression with respect to time, we have:

$$\frac{dE}{dt} = m \left(\frac{dx}{dt} \right) \times \frac{d^2x}{dt^2} + kx \frac{dx}{dt}$$

This leads to:

$$\frac{dE}{dt} = \left(\frac{dx}{dt} \right) \left[m \frac{d^2x}{dt^2} + kx \right] \qquad [1.298]$$

Applying Newton's second law, we get:

$$F = ma \Rightarrow -kx = m \frac{d^2x}{dt^2}$$

This finally leads to:

$$m\frac{d^2x}{dt^2}+kx=0 \tag{1.299}$$

Considering [1.298] and [1.299], we finally have: $dE/dt = 0 \Rightarrow E = C_{st}$: the classical harmonic oscillator is definitely a conservative system (*in fact, all fluid friction is assumed zero*).

(3) *Behavior of the quantum harmonic oscillator*

(3.1) *Proof*

The stationary Schrödinger equation of the quantum oscillator has the following form:

$$\left[-\frac{\hbar^2}{2m}\frac{d^2}{dx^2}+V(x)\right]\Phi(x)=E\Phi(x) \tag{1.300}$$

Using [1.297], after arrangement equation [1.300] becomes:

$$-\frac{d^2\Phi(x)}{dx^2}+\frac{mk}{\hbar^2}x^2\Phi(x)=\frac{2mE}{\hbar^2}\Phi(x) \tag{1.301}$$

The angular frequency of the oscillator described by equation [1.299] verifies the well-known relation $\omega^2 = k/m \Rightarrow m = k/\omega^2$. Equation [1.301] can be written as:

$$-\frac{d^2\Phi(x)}{dx^2}+\left(\frac{kx^2}{\hbar\omega}\right)\frac{k}{\hbar\omega}\Phi(x)=\left(\frac{2E}{\hbar\omega}\right)\frac{k}{\hbar\omega}\Phi(x) \tag{1.302}$$

We introduce the dimensionless quantities α and q such that:

$$\begin{cases}\alpha=\dfrac{2E}{\hbar\omega}\\[2ex] q^2=\dfrac{k}{\hbar\omega}x^2 \Rightarrow x=\sqrt{\dfrac{\hbar\omega}{k}}\,q\end{cases} \tag{1.303}$$

We now express the second derivative of the wave function with respect to the variable x. Taking the second relation of [1.303] into account, we have:

$$\frac{d\Phi(q)}{dq}=\frac{d\Phi(x)}{dx}\frac{dx}{dq}=\sqrt{\frac{\hbar\omega}{k}}\frac{d\Phi(x)}{dx}$$

Using this equality, we have:

$$\frac{d^2\Phi(q)}{dq^2} = \sqrt{\frac{\hbar\omega}{k}} \frac{d}{dx}\left(\frac{d\Phi(x)}{dx}\right) \times \frac{dx}{dq}$$

Hence:

$$\frac{d^2\Phi(q)}{dq^2} = \frac{\hbar\omega}{k} \frac{d^2\Phi(x)}{dx^2}$$

We finally have:

$$\frac{d^2\Phi(x)}{dx^2} = \frac{k}{\hbar\omega} \frac{d^2\Phi(q)}{dq^2} \qquad [1.304]$$

Using [1.303] and [1.304], equation [1.302] can be written after simplification:

$$-\frac{d^2\Phi(q)}{dq^2} + q^2\Phi(q) = \alpha\,\Phi(q) \qquad [1.305]$$

(3.2) *Proof*

The wave function of the ground state has the form $\Phi_0(q) = \exp(\beta q^2)$, where β is a constant. Using this solution, equation [1.305] is written as:

$$\frac{d^2\Phi(q)}{dq^2} = \left(2\beta + 4q^2\beta^2\right)\Phi(q)$$

This leads to:

$$\left(1 - 4\beta^2\right)q^2 - 2\beta = \alpha \qquad [1.306]$$

(3.3) *Possible values*

Equation [1.306] only has a solution if it is identically verified with respect to variable q. Hence: $(1 - 4\beta^2) = 0$ and $-2\beta = \alpha$. We then have $\beta = \pm 1/2$. The value $\beta = 1/2$ gives a wave function $\Phi_0(q) = \exp(q^2/2)$, which is divergent when $q \to \infty$. The value $\beta = -½$ is retained.

(3.4) *Expression of the ground state energy*

It is known that $-2\beta = \alpha$. Since we know that $\beta = -1/2$, $\alpha = 1$. If the energy of the ground state of the quantum harmonic oscillator is designated by E_0, the first relation of system [1.303] thus becomes:

$$E_0 = \frac{\hbar\omega}{2} \quad [1.307]$$

1.8.11. *Solution 11 – Quantized energy of the harmonic oscillator*

The dimensionless quantities are:

$$q = \sqrt{\frac{m\omega}{\hbar}}x\,;\ \varepsilon = \frac{2E}{\hbar\omega} \quad [1.308]$$

(1) *Proof*

The stationary Schrödinger equation of the *quantum harmonic oscillator* is given by [1.305] where α must be replaced by ε. We obtain:

$$\frac{d^2\Phi(q)}{dq^2} + \left(\varepsilon - q^2\right)\Phi(q) = 0 \quad [1.309]$$

(2) *Differential equation*

The solution to equation [1.309] has the form:

$$\Phi(q) = Au(q)e^{-q^2/2} \quad [1.310]$$

where A is a constant and the function $u\,(q)$ is a complete series of powers of q:

$$u(q) = \sum_{k=0}^{\infty} a_k\, q^k \quad [1.311]$$

Differentiating twice the wave function [1.310], we have:

$$\begin{cases} \dfrac{d\Phi(q)}{dq} = A\left[\dfrac{du(q)}{dq} - qu(q)\right]e^{-q^2/2} \\ \dfrac{d^2\Phi(q)}{dq^2} = A\left[\dfrac{d^2u(q)}{dq^2} - q\dfrac{du(q)}{dq} - u(q)\right]e^{-q^2/2} - qA\left[\dfrac{du(q)}{dq} - qu(q)\right]e^{-q^2/2} \end{cases}$$

The second of these equations can be written as:

$$\frac{d^2\Phi(q)}{dq^2} = \left[\frac{d^2u(q)}{dq^2} - 2q\frac{du(q)}{dq} + (q^2-1)u(q)\right]e^{-q^2/2}$$

If we insert this result in [1.309], after arrangement we get:

$$\frac{d^2u(q)}{dq^2} - 2q\,\frac{du(q)}{dq} + (\varepsilon - 1)\,u(q) = 0 \qquad [1.312]$$

(3) *Expression*

Using [1.311], we have:

$$\begin{cases} \dfrac{du(q)}{dq} = \displaystyle\sum_{k=0}^{\infty} ka_k\, q^{k-1} \\ \dfrac{d^2u(q)}{dq^2} = \displaystyle\sum_{k=0}^{\infty} k(k-1)a_k\, q^{k-2} \end{cases} \qquad [1.313]$$

Using [1.313], equation [1.312] can be written as:

$$\sum_{k=0}^{\infty} k(k-1)a_k\, q^{k-2} - 2q\sum_{k=0}^{\infty} ka_k\, q^{k-1} + (\varepsilon - 1)\sum_{k=0}^{\infty} a_k\, q^k = 0$$

Hence:

$$\sum_{k=0}^{\infty} k(k-1)a_k\, q^{k-2} + \sum_{k=0}^{\infty}(\varepsilon - 2k - 1)a_k\, q^k = 0 \qquad [1.314]$$

Identifying the terms of the same power in q^k, we find (it suffices to replace k by $k + 2$ in the first term of equation [1.314]):

$$(k+2)(k+1)a_{k+2} + (\varepsilon - 2k - 1)a_k = 0$$

This finally leads to:

$$a_{k+2} = \frac{(2k+1-\varepsilon)}{(k+2)(k+1)} a_k \qquad [1.315]$$

(4) *Quantization of energy*

The wave function [1.310] is finite or convergent for all the values of q (including $q \to \pm\infty$), provided that the series [1.315] stops for a certain value n of the integer k. This cut-off condition requires the coefficient $a_{n+2} = 0$. Or, according to the series [1.315]:

$$2n + 1 - \varepsilon = 0 \Rightarrow \varepsilon = 2n + 1 \qquad [1.316]$$

Using [1.309], after arrangement we have:

$$E_n = \hbar\omega\left(n + \frac{1}{2}\right) \qquad [1.317]$$

The ground state energy of the oscillator is obtained for the minimal value $n = 0$. This yields $E_0 = \hbar\omega/2$ according to [1.307].

(5) *Potential energy variation curve*

Expression [1.297] shows that the representative curve of the potential energy $V(x)$ is a parabola (Figure 1.22).

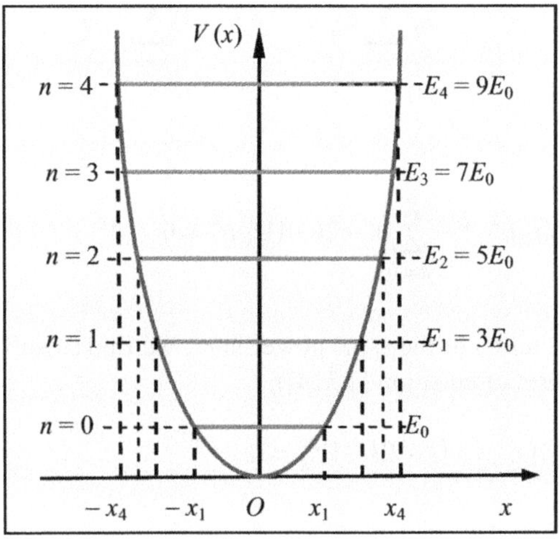

Figure 1.22. *Curve of the variation of the potential energy of a harmonic oscillator with position x*

From the perspective of classical mechanics, the particle is subjected to the opposing spring force $F = -kx$ and oscillates around its equilibrium position at the origin O of abscissa $x_0 = 0$.

In quantum mechanics, the notion of opposing spring force is devoid of meaning. The harmonic oscillator is defined as a particle (atoms in vibration motion in a polar molecule such as HCl (see the solution in section 1.8.12), thermal agitation of atoms or nuclei in a crystal lattice, etc.) whose potential energy is given by the expression [1.297]. The particle is then confined in a parabolic potential well (Figure 1.22) in which the energy levels are discrete and equidistant.

(6) *Classical and quantum predictions*

According to Figure 1.22, the harmonic oscillator is at rest at its equilibrium position in $x_0 = 0$. From a classical point of view, its energy is therefore zero. On the other hand, from the quantum perspective, the ground state energy of the oscillator is non-zero (given the uncertainty principle) and is equal to $E_0 = \hbar\omega/2$. This means there is an essential difference between the classical and quantum predictions of the behavior of the harmonic oscillator.

NOTE.– At absolute zero temperature, the oscillator does not oscillate. The energy at absolute zero is thus a purely quantum quantity and it corresponds to the oscillation energy of the quantum vacuum. Indeed, in quantum electrodynamics, any electromagnetic field can be assimilated to a set of independent *virtual harmonic oscillators* for which the principal quantum number n in the quantized expression [1.317] refers to the number of photons. Hence, the situation of the ground state for which $n = 0$ corresponds to the absence of photons, which is a *photonic vacuum* in the enclosure initially filled by a radiation field (see Appendix 1, Volume 1). The residual energy E_0 is thus called *zero-oscillation energy* ($T = 0$ K) *of the photonic vacuum.*

(7) *Ground state energy*

From a classical point of view, the total energy of the harmonic oscillator is given by the sum of its kinetic and potential energies. Knowing that the potential energy is defined up to an additive constant, we obtain:

$$E = \frac{1}{2}mv^2 + \frac{1}{2}kx^2 + C^{st} \qquad [1.318]$$

If the particle is at rest in $x = 0$, then the total energy is equal to C^{st}. We can intuitively consider $C^{st} \equiv E_0 = \hbar\omega/2$. This reconciles the predictions of classical mechanics and those of quantum mechanics. Caution is however recommended, as there is no classical law that makes it possible to prove that $E_0 = \hbar\omega/2$. The energy E_0 is a purely quantum property of the harmonic oscillator studied in this exercise.

1.8.12. *Solution 12 – HCl molecule assimilated to a linear oscillator*

The potential energy of a HCl dipole has the following form:

$$V(x) = \frac{1}{2}k(x-a)^2 \qquad [1.319]$$

The wave function $\Phi_0(x)$ of the ground state and that of the first excited state $\Phi_1(x)$ are given by the following expressions:

$$\Phi_0(x) = A_0 e^{-\alpha^2(x-a)^2/2}; \Phi_1(x) = A_1(x-a)e^{-\beta^2(x-a)^2/2} \qquad [1.320]$$

In relations [1.320], A_0 and A_1 are normalization constants, and α and β are strictly positive constants.

(1) *Schrödinger's equation of stationary states, proof*

– *Schrödinger's equation*

The Schrödinger equation of vibration stationary states of the HCl molecule is given by the following expression:

$$\left[-\frac{\hbar^2}{2m}\frac{d^2}{dx^2} + V(x)\right]\Phi(x) = E\Phi(x)$$

Using [1.320], we obtain:

$$\frac{d^2\Phi(x)}{dx^2} + \frac{2m}{\hbar^2}\left[E - \frac{1}{2}k(x-a)^2\right]\Phi(x) = 0 \qquad [1.321]$$

– *Relation between α and β*

Let us determine the second derivatives of the wave functions [1.321] with respect to variable x. We obtain:

$$\begin{cases} \dfrac{d\Phi_0(x)}{dx} = \left[-\alpha^2(x-a)\right]\Phi_0(x) \\ \dfrac{d^2\Phi_0(x)}{dx^2} = \alpha^2\left[\alpha^2(x-a)^2 - 1\right]\Phi_0(x) \end{cases} \qquad [1.322]$$

$$\begin{cases} \dfrac{d\Phi_1(x)}{dx} = \left[1-\beta^2(x-a)^2\right] A_1 e^{-\beta^2(x-a)^2/2} \\ \dfrac{d^2\Phi_1(x)}{dx^2} = \beta^2\left[-3+\beta^2(x-a)^2\right]\Phi_1(x) \end{cases} \quad [1.323]$$

Taking [1.322] into account, after arrangement equation [1.321] gives:

$$\left[\alpha^4(x-a)^2-\alpha^2\right]+\left[\frac{2mE_0}{\hbar^2}-\frac{km}{\hbar^2}(x-a)^2\right]=0$$

Hence:

$$\left(\alpha^4-\frac{km}{\hbar^2}\right)(x-a)^2+\left(\frac{2mE_0}{\hbar^2}-\alpha^2\right)=0 \quad [1.324]$$

This equation is verified in all cases if:

$$\begin{cases} \left(\alpha^4-\dfrac{km}{\hbar^2}\right)=0 \Rightarrow \alpha^2=\sqrt{\dfrac{km}{\hbar^2}} \\ \left(\dfrac{2mE_0}{\hbar^2}-\alpha^2\right)=0 \Rightarrow \alpha=\sqrt{\dfrac{2mE_0}{\hbar^2}} \end{cases} \quad [1.325]$$

Similarly, using [1.323], after arrangement equation [1.321] gives :

$$\left(\beta^4-\frac{km}{\hbar^2}\right)(x-a)^2+\left(\frac{2mE_1}{\hbar^2}-3\beta^2\right)=0 \quad [1.326]$$

As previously explained, this equation is verified in all cases if:

$$\begin{cases} \left(\beta^4-\dfrac{km}{\hbar^2}\right)=0 \Rightarrow \beta^2=\sqrt{\dfrac{km}{\hbar^2}} \\ \left(\dfrac{2mE_1}{\hbar^2}-3\beta^2\right)=0 \Rightarrow \beta=\sqrt{\dfrac{2mE}{3\hbar^2}} \end{cases} \quad [1.327]$$

Comparing the first relations in [1.325] and [1.327], we have: $\alpha=\beta$.

(2) *Expressions of energies*

For the studied harmonic oscillator, the angular frequency of the oscillations verifies the relation $\omega^2 = k/m$. Hence, we have: $km = \omega^2 m^2$. Therefore, the first relation of equations [1.325] gives:

$$\alpha^2 = \sqrt{\frac{m^2\omega^2}{\hbar^2}} = \frac{m\omega}{\hbar} \quad [1.328]$$

Since $\alpha = \beta$, we have:

$$\alpha = \beta = \sqrt{\frac{m\omega}{\hbar}} \quad [1.329]$$

The expressions of the respective energies E_0 and E_1 of the ground state and of the first excited state of the HCl molecule are automatically expressed considering the second relations of equations [1.325] and [1.327]. Hence:

$$\begin{cases} \alpha = \sqrt{\dfrac{2mE_0}{\hbar^2}} = \sqrt{\dfrac{m\omega}{\hbar}} \\ \beta = \sqrt{\dfrac{2mE_1}{3\hbar^2}} = \sqrt{\dfrac{m\omega}{\hbar}} \end{cases}$$

From the previous relations, we deduce:

$$\begin{cases} E_0 = \dfrac{\hbar\omega}{2} \\ E_1 = \dfrac{3}{2}\hbar\omega \end{cases} \quad [1.330]$$

Results [1.330] can be directly obtained from the general expression [1.317] of the energy E_n considering $n = 0$ and $n = 1$, respectively.

(3) *Expressions of the normed wave functions*

Family of integrals:

$$I_p = \int_0^\infty x^p e^{-\rho x^2} dx\,; I_p = \frac{p-1}{2\rho} I_{p-2}\,; I_0 = \frac{1}{2}\sqrt{\frac{\pi}{\rho}}\,;\; I_1 = \frac{1}{2\rho} \quad [1.331]$$

Constants A_0 and A_1 are determined by the normalization condition satisfied by the wave function. Or, using [1.320]:

$$\int_{-\infty}^{+\infty} |\Phi(x)|^2 \, dx = 1 \Rightarrow \begin{cases} |A_0|^2 \int_{-\infty}^{+\infty} e^{-\alpha^2 (x-a)^2} dx = 1 \\ |A_1|^2 \int_{-\infty}^{+\infty} (x-a)^2 e^{-\alpha^2 (x-a)^2} dx = 1 \end{cases} \quad [1.332]$$

Let us make the change of variable: $y = x - a$. Considering that the wave functions are even, equations [1.332] are written as:

$$\begin{cases} 2|A_0|^2 \int_0^{+\infty} e^{-\alpha^2 y^2} dy = 1 \\ 2|A_1|^2 \int_0^{+\infty} y^2 e^{-\alpha^2 y^2} dy = 1 \end{cases}$$

After integration, we have:

$$\begin{cases} 2|A_0|^2 I_0 = 1 \Rightarrow |A_0|^2 = \dfrac{1}{2I_0} \\ 2|A_1|^2 I_2 = 1 \Rightarrow |A_1|^2 = \dfrac{1}{2I_2} = \dfrac{\rho}{I_0} \end{cases}$$

where $\rho = \alpha^2$. Using [1.329] and [1.331], if A_0 and A_1 are real, we have:

$$|A_0|^2 = \frac{\alpha}{\sqrt{\pi}} ; |A_1|^2 = 2\frac{\alpha^3}{\sqrt{\pi}}$$

$$\begin{cases} A_0 = \left(\dfrac{m\omega}{\pi\hbar}\right)^{1/4} \\ A_1 = \left[\dfrac{4}{\pi}\left(\dfrac{m\omega}{\hbar}\right)^3\right]^{1/4} \end{cases} \quad [1.333]$$

Using [1.333], the normed wave functions are written according to [1.320]:

$$\Phi_0(x) = \left(\frac{m\omega}{\pi\hbar}\right)^{\frac{1}{4}} e^{-\frac{1}{2}\frac{m\omega}{\hbar}(x-a)^2} \qquad [1.334]$$

$$\Phi_1(x) = \left[\frac{4}{\pi}\left(\frac{m\omega}{\hbar}\right)^3\right]^{\frac{1}{4}} (x-a)e^{-\beta^2(x-a)^2/2} \qquad [1.335]$$

NOTE.– A detailed study of the properties of the classical harmonic oscillator is provided in the references [COH 77, GRI 95, MAR 00]. The wave functions associated with the stationary states of the harmonic oscillator are given by the general expression:

$$\Phi_n(x) = \left[\frac{1}{2^n n!}\left(\frac{\hbar}{m\omega}\right)^n\right]^{\frac{1}{2}} \left(\frac{m\omega}{\hbar}\right)^{\frac{1}{4}} \times \left[\frac{m\omega}{\hbar}x - \frac{d}{dx}\right]^n e^{-\frac{1}{2}\frac{m\omega}{\hbar}x^2} \qquad [1.336]$$

Thus we find the expressions [1.334] and [1.335] by changing in the general wave function [1.336] x into $x - a$ for $n = 0$ and $n = 1$, respectively.

1.8.13. *Solution 13 – Quantized energy of hydrogen-like systems*

Schrödinger equation describing the evolution of the radial function of hydrogen-like systems:

$$\frac{d^2\Phi(r)}{dr^2} + \frac{2}{r}\frac{d\Phi(r)}{dr} + \left(\frac{\delta}{r} - \varepsilon^2\right)\Phi(r) = 0 \qquad [1.337]$$

In this equation:

$$\varepsilon^2 = -\frac{2mE}{\hbar^2}; \delta = 2\frac{kZme^2}{\hbar^2} \qquad [1.338]$$

$k = 1/4\pi\varepsilon_0$

$$\Phi(r) = \frac{\chi(r)}{r}e^{-\varepsilon r}\ ; \chi(r) = \sum_{k=\nu}^{\infty} a_k r^k \qquad [1.339]$$

where ν is a positive integer.

(1) *Differential equation*

Using [1.339], the first derivative of function $\Phi(r)$ is:

$$\frac{d\Phi(r)}{dr}=\left(-\frac{\chi}{r^2}+\frac{1}{r}\frac{d\chi}{dr}-\frac{\varepsilon\chi}{r}\right)e^{-\varepsilon r} \quad [1.340]$$

hence:

$$\frac{2}{r}\frac{d\Phi(r)}{dr}=\left(-\frac{2\chi}{r^3}+\frac{2}{r^2}\frac{d\chi}{dr}-\frac{2\varepsilon\,\chi}{r^2}\right)e^{-\varepsilon r} \quad [1.341]$$

The second derivative is:

$$\frac{d^2\Phi(r)}{dr^2}=\left(\frac{2\chi}{r^3}-\frac{2}{r^2}\frac{d\chi}{dr}+\frac{2\varepsilon\,\chi}{r^2}\right)e^{-\varepsilon r}+\left(\frac{1}{r}\frac{d^2\chi}{dr^2}-\frac{2\varepsilon}{r}\frac{d\chi}{dr}+\frac{\varepsilon^2\,\chi}{r}\right)e^{-\varepsilon r} \quad [1.342]$$

Moreover, using [1.340], we have:

$$\left(\frac{\delta}{r}-\varepsilon^2\right)\Phi(r)=\left(\frac{\delta\,\chi}{r^2}-\varepsilon^2\frac{\chi}{r}\right)e^{-\varepsilon r} \quad [1.343]$$

Summing equations [1.341], [1.342] and [1.343], we find:

$$\frac{d^2\chi(r)}{dr^2}-2\varepsilon\frac{d\chi(r)}{dr}+\frac{\delta}{r}\chi(r)=0 \quad [1.344]$$

(2.1) *Proof*

We express the first and second derivatives of function $\chi(r)$ using [1.339]. This leads to:

$$\begin{cases}\dfrac{d\chi}{dr}=\sum\limits_{k=\nu}^{\infty}ka_k\,r^{k-1}\\[2ex]\dfrac{d^2\chi}{dr^2}=\sum\limits_{k=\nu}^{\infty}k(k-1)a_k\,r^{k-2}\end{cases} \quad [1.345]$$

Using [1.340], the differential equation [1.344] gives:

$$\sum_{k=\nu}^{\infty} k(k-1)a_k r^{k-2} - 2\varepsilon \sum_{k=\nu}^{\infty} ka_k r^{k-1} + \frac{\delta}{r}\sum_{k=\nu}^{\infty} a_k r^k = 0 \qquad [1.346]$$

Two cases can be distinguished.

– *First case: k = ν.*

Using equation [1.346], we have:

$$\nu(\nu-1)\, a_\nu r^{\nu-2} + (\delta - 2\varepsilon\nu)\, a_\nu r^{\nu-1} = 0$$

This equation is verified in all cases if:

$$\begin{cases} \nu(\nu-1) = 0 \\ (\delta - 2\varepsilon\nu) = 0 \end{cases} \qquad [1.347]$$

– *Second case: k ≠ ν.*

Expanding equation [1.346], we obtain the terms of the same power in r^k. The identification of these terms leads to:

$$k(k+1)\, a_{k+1} r^{k-1} - 2\varepsilon k\, a_k r^{k-1} + \delta\, a_k r^{k-1} = 0$$

Hence:

$$\left[k(k+1)\, a_{k+1} - 2\varepsilon k\, a_k + \delta\, a_k\right] r^{k-1} = 0 \qquad [1.348]$$

Considering [1.348] and the results [1.347], we finally get:

$$\begin{cases} \nu(\nu-1) & = 0 \\ k(k+1)a_{k+1} - 2\varepsilon k\, a_k + \delta\, a_k & = 0 \end{cases} \qquad [1.349]$$

(2.2) *Possible values of ν*

The first solution of [1.348] leads to $\nu = 0$ or $\nu = 1$. For $\nu = 0$, the series [1.339] contains one constant term and is written as:

$$\chi(r) = a_0 + \sum_{k\neq 0}^{\infty} a_k r^k \qquad [1.350]$$

The wave function [1.346] has the form:

$$\Phi(r) = \frac{a_0}{r} + \frac{1}{r}\sum_{k\neq 0}^{\infty} a_k r^k \times e^{-\varepsilon r} \qquad [1.351]$$

Expression [1.351] shows that for $\nu = 0$, the wave function has a term that tends toward infinity when $r = 0$ (a_0 is not zero). Since the wave function must be bounded, the only solution to be retained is $\nu = 1$. For this solution, we have:

$$\Phi(r) = a_1 e^{-\varepsilon r} \qquad [1.352]$$

This corresponds to the solution characterizing the ground state of the hydrogen-like system. The only difference between this solution and [1.144] resides in the factor a_1, which is determined from the normalization condition that must be met by the wave function $\Phi(r)$.

(3) *Expression, asymptotic behavior*

Using the second equation of results [1.349], we have:

$$\frac{a_{k+1}}{a_k} = \frac{2\varepsilon k - \delta}{k(k+1)} \qquad [1.353]$$

At infinity ($k \to \infty$), the ratio a_{k+1}/a_k tends toward:

$$\frac{a_{k+1}}{a_k} \to \frac{2\varepsilon}{k+1} \qquad [1.354]$$

(4) *Comparison*

The expansion in complete series of the function $e^{2\varepsilon r}$ can be written as follows:

$$e^{2\varepsilon r} = \sum_{k=0}^{\infty} b_k r^k = \sum_{k=0}^{\infty} \frac{1}{k!}(2\varepsilon)^k r^k \qquad [1.355]$$

The coefficients b_k of this expansion are given by the relation:

$$b_k = \frac{1}{k!}(2\varepsilon r)^k \qquad [1.356]$$

Using this expression, the ratio b_{k+1}/b_k can be written as follows:

$$\frac{b_{k+1}}{b_k} = \frac{2\varepsilon}{k+1} \qquad [1.357]$$

This corresponds to the convergence limit [1.354] of the ratio $a_{k+1}/a_{k.}$. Consequently, at infinity the series [1.339] behaves as the function $e^{2\varepsilon r}$. It can be noted that the wave function [1.339] diverges if we consider $\chi\ (r) = e^{2\varepsilon r}$. This justifies the cut-off condition.

CONCLUSION.– The series [1.339] is therefore convergent. It presents a *cut-off*, meaning that it stops for a certain well-determined integer value of k.

(5) *Expression of the quantized energy*

The cut-off condition of the series [1.339] is obtained for $k = n$, where n is a positive integer taking the values 1, 2, 3… (it is worth recalling that the smallest value of k is equal to $k_{\min} = \nu = 1$). Hence, the series stops for $k = n$. It follows from this that the coefficient a_{k+1} in [1.353] is zero. Hence: $2\varepsilon n = \delta$. Using [1.338], we have:

$$\frac{2mE}{\hbar^2} = -\left(\frac{Zme^2}{\hbar^2}\right)^2 \times \frac{1}{n^2}$$

Arranging this relation, we finally have:

$$E_n = -\frac{Z^2 me^4}{2\hbar^2 n^2} \qquad [1.358]$$

1.8.14. *Solution 14 – Line integral of the probability current density vector, Bohr's magneton*

Probability current density:

$$\vec{J}(\vec{r},t) = \frac{\hbar}{2mi}\left[\Psi^*\vec{\nabla}\Psi - \Psi\vec{\nabla}\Psi^*\right] \qquad [1.359]$$

Wave function:

$$\Psi\ (r,\theta,\ \varphi) = \mathrm{R}\ (r) \times \Theta\ (\theta) \times \Phi\ (\varphi);\ \Phi(\varphi) = \exp\ (i\, m_\ell\ \varphi) \qquad [1.360]$$

$$(\vec{\nabla}\Psi)_r = \frac{\partial \Psi}{\partial r}\,;\,(\vec{\nabla}\Psi)_\theta = \frac{1}{r}\frac{\partial \Psi}{\partial \theta}\,;(\vec{\nabla}\Psi)_\varphi = \frac{1}{r\sin\theta}\frac{\partial \Psi}{\partial \varphi} \qquad [1.361]$$

(1) *Components of the probability current density vector*

According to expression [1.359], the components J_r and J_θ involve the real parts R (r) and Θ (θ), respectively: these components are therefore zero.

The component J_φ is written as:

$$J_\varphi = \frac{\hbar}{2mi}\left[\Psi * \nabla_\varphi \Psi) - \Psi \nabla_\varphi \Psi *\right] \qquad [1.362]$$

Taking [1.360] and [1.361] into account, we have:

$$\begin{cases} \Psi * \nabla_\varphi \Psi = \Psi * \dfrac{im_\ell}{r\sin\theta}\Psi \\ \Psi \nabla_\varphi \Psi * = \Psi \dfrac{-im_\ell}{r\sin\theta}\Psi * \end{cases} \qquad [1.363]$$

Expression [1.362] is then written as follows:

$$J_\varphi = \frac{\hbar}{2mi} \times \frac{im_\ell}{r\sin\theta} 2\Psi * \Psi$$

This leads to:

$$J_\varphi = \frac{\hbar}{m} m_\ell \times \frac{\Psi * \Psi}{r\sin\theta} \qquad [1.364]$$

(2) *Expression of intensity*

Intensity dI_φ of the current through $d\sigma$ (the flux of the current density vector through the elementary surface $d\sigma$) is defined by the following relation:

$$dI_\varphi = j_\varphi\, d\sigma \qquad [1.365]$$

By definition, the component of the current density vector is $j_\varphi = -eJ_\varphi$. Using [1.364] and [1.365], we get:

$$j_\varphi = -\frac{e\hbar}{m} m_\ell \frac{\Psi * \Psi}{r\sin\theta}$$

Hence, we have:

$$dI_\varphi = -\frac{e\hbar}{m} m_\ell \frac{\Psi^*\Psi}{r\sin\theta} d\sigma \qquad [1.366]$$

(3) *Expression of the elementary magnetic moment*

The *magnetic moment* is the product of the intensity and the area delimited by the current. Knowing that the delimited area is $S = \pi\rho^2 = \pi(r\sin\theta)^2$, we have:

$$dM_\ell = dI_\varphi S \Rightarrow dM_\ell = dI_\varphi \pi (r\sin\theta)^2 \qquad [1.367]$$

Inserting [1.366] into [1.367], we find:

$$dM_\ell = -\frac{e\hbar}{m} m_\ell \frac{\Psi^*\Psi}{r\sin\theta} \pi (r\sin\theta)^2 d\sigma$$

This leads to:

$$dM_\ell = -\frac{e\hbar}{2m} m_\ell \Psi^*\Psi \, d\tau \qquad [1.368]$$

In relation [1.368], the elementary volume is $d\tau = 2\pi r \sin\theta d\sigma$.

(4) *Orbital magnetic moment, Bohr's magneton*

– *Magnetic moment*

Integrating [1.368] on all the current tubes (which amounts to integration throughout the space), we obtain:

$$M_\ell = \int dM_\ell = -\frac{e\hbar}{2m} m_\ell \int \Psi^*\Psi \, d\tau \qquad [1.369]$$

Knowing that the wave function is normed:

$$M_\ell = -\frac{e\hbar}{2m} m_\ell \qquad [1.370]$$

– *Bohr's magneton*

Expression [1.370] shows that the orbital magnetic moment is the product of the dimensionless magnetic quantum number m_ℓ and a magnetic moment denoted μ_B, known as Bohr's magneton. By definition, $\mu_B = -M_\ell/m_\ell$. Hence:

$$\mu_B = \frac{e\hbar}{2m} \quad [1.371]$$

Numerical expression:

$$\mu_B = 9.274 \times 10^{-24}\ \text{Am}^2.$$

1.8.15. *Solution 15 – Schrödinger's equation in the presence of a magnetic field, Zeeman–Lorentz triplet*

In the absence of a field of external forces, the stationary Schrödinger equation describing the evolution of the state of a free particle is written as:

$$\Delta\Psi + \frac{2m}{\hbar^2}(E - V)\Psi = 0 \quad [1.372]$$

In the presence of a magnetic field, equation [1.372] takes the form:

$$\Delta\Psi - i\frac{2e}{\hbar}\vec{A}.\vec{\nabla}\Psi + \frac{2m}{\hbar^2}(E - V)\Psi = 0 \quad [1.373]$$

The wave function of the hydrogen-like system $\Psi\ (r, \theta, \varphi) = \text{R}\ (r) \times \Theta\ (\theta) \times \Phi\ (\varphi)$, the imaginary angular part $\Phi(\varphi) = \exp\ (i m_\ell\ \varphi)$. The relations between Cartesian coordinates and spherical coordinates are:

$$x = r \sin\theta \cos\varphi;\ y = r \sin\theta \sin\varphi;\ z = r \cos\theta \quad [1.374]$$

(1) *Proof*

The wave function depends on x, y, z through r. Hence:

$$\frac{\partial\Psi}{\partial\varphi} = \frac{\partial\Psi}{\partial x}\frac{\partial x}{\partial\varphi} + \frac{\partial\Psi}{\partial y}\frac{\partial y}{\partial\varphi} + \frac{\partial\Psi}{\partial z}\frac{\partial z}{\partial\varphi} \quad [1.375]$$

Taking [1.374] into account, expression [1.375] leads to:

$$\frac{\partial\Psi}{\partial\varphi} = -r\sin\theta\sin\varphi\frac{\partial\Psi}{\partial x} + r\sin\theta\cos\varphi\frac{\partial\Psi}{\partial y}$$

Therefore, after arrangement we have:

$$\frac{\partial \Psi}{\partial \varphi} = x\frac{\partial \Psi}{\partial y} - y\frac{\partial \Psi}{\partial x} \qquad [1.376]$$

(2) *Deduction of the magnetic field*

The vector potential $\vec{A}$ is given by the relation of definition $\vec{B} = \vec{\nabla} \wedge \vec{A}$. To answer the question, it is sufficient to prove that the direction of the magnetic field is that of the Oz axis, hence: $B = B_z$. Let us consider the following relations:

$$A_x = -\frac{1}{2}B\,y; \; A_y = \frac{1}{2}B\,x \text{ and } A_z = 0 \qquad [1.377]$$

We have:

$$\begin{cases} B_x = \left(\vec{\nabla} \wedge \vec{A}\right)_x = \dfrac{\partial A_z}{\partial y} - \dfrac{\partial A_y}{\partial z} = 0 \\ B_y = \left(\vec{\nabla} \wedge \vec{A}\right)_y = \dfrac{\partial A_x}{\partial z} - \dfrac{\partial A_z}{\partial x} = 0 \\ B_z = \left(\vec{\nabla} \wedge \vec{A}\right)_z = \dfrac{\partial A_y}{\partial x} - \dfrac{\partial A_x}{\partial y} = \dfrac{B}{2} + \dfrac{B}{2} = B \end{cases}$$

In conclusion, the magnetic field derives from the vector potential whose coordinates are given by relations [1.377].

(3) *Proof*

Let us determine the quantity $\vec{A}.\vec{\nabla}\Psi$ in the expression [1.373]. We have:

$$\vec{A} \cdot \vec{\nabla}\Psi = A_x\frac{\partial \Psi}{\partial x} + A_y\frac{\partial \Psi}{\partial y} + A_z\frac{\partial \Psi}{\partial z} \qquad [1.378]$$

Taking [1.376] into account, we have:

$$\vec{A} \cdot \vec{\nabla}\Psi = \frac{1}{2}B\left(x\frac{\partial \Psi}{\partial y} - y\frac{\partial \Psi}{\partial x}\right) \qquad [1.379]$$

Using [1.376] and [1.379], we obtain:

$$\vec{A}\cdot\vec{\nabla}\Psi = \frac{1}{2}B\frac{\partial\Psi}{\partial\varphi}$$

This leads to:

$$-i\frac{2e}{\hbar}\vec{A}\cdot\vec{\nabla}\Psi = -i\frac{e}{\hbar}B\frac{\partial\Psi}{\partial\varphi} \qquad [1.380]$$

Knowing that $\Psi\,(r,\theta,\,\varphi) = \mathrm{R}\,(r)\times\Theta\,(\theta)\times\exp\,(i\,m_\ell\,\varphi)$, [1.380] is finally written as:

$$-i\frac{2e}{\hbar}\vec{A}\cdot\vec{\nabla}\Psi = \frac{e}{\hbar}Bm_\ell\Psi \qquad [1.381]$$

(4) *Schrödinger's equation*

Inserting [1.381] into equation [1.373], we find:

$$\Delta\Psi + \frac{e}{\hbar}Bm_\ell\Psi + \frac{2m}{\hbar^2}(E-U)\Psi = 0 \qquad [1.382]$$

The Schrödinger equation in the presence of a magnetic field [1.382] can be written as:

$$\Delta\Psi + \Psi + \frac{2m}{\hbar^2}\left(E + m_\ell\frac{e\hbar}{2m}B - V\right)\Psi = 0$$

Writing this equation in the form of the Schrödinger equation [1.372], we have:

$$\Delta\Psi + \frac{2m}{\hbar^2}(E' - V)\Psi = 0 \qquad [1.383]$$

In equation [1.383], E' is the total energy of the studied hydrogen-like system in the presence of a magnetic field with:

$$E' = E + m_\ell\frac{e\hbar}{2m}B \qquad [1.384]$$

(5) *Expression of total energy, Zeeman–Lorentz triplet*

– *Expression of total energy*

For a hydrogen-like system with potential energy $V(r) = -Ze^2/r$ and total energy E, the solution to the stationary Schrödinger equation [1.372] gives the discrete values [1.359] of the energy E, which are the following:

$$E_n = -\frac{Z^2 me^4}{2\hbar^2 n^2}$$

Consequently, the solution [1.383] of the Schrödinger equation in the presence of the magnetic field can be written as follows:

$$E_n' = E_n + m_\ell \frac{e\hbar}{2m} B \quad [1.385]$$

Result [1.385] expresses the quantization of the total energy of the hydrogen-like system, which now depends on the angular momentum quantum number.

– *Zeeman–Lorentz triplet*

During a transition between states $\ell = 0$ and $\ell = 1$, the energy varies by the quantity:

$$\Delta E_n' = \Delta E_n + \frac{e\hbar}{2m} B \Delta m_\ell \quad [1.386]$$

Knowing that $\Delta E = \hbar\omega$ and Larmor's frequency is $\Omega = eB/2m$, we have:

$$\omega = \omega_0 + \frac{eB}{2m} \Delta m_\ell \Rightarrow \omega = \omega_0 + \Omega \Delta m_\ell \quad [1.387]$$

The result [3.187] effectively gives the Zeeman–Lorentz triplet [3.188].

1.8.16. *Solution 16 – Deduction of Schrödinger's equation from the De Broglie relation*

(1) *Differential equation*

Let us recall the propagation differential equation:

$$\frac{\partial^2 \Psi}{\partial q^2} = \frac{1}{c^2} \frac{\partial^2 \Psi}{\partial t^2} \quad [1.388]$$

For an electromagnetic wave, the frequency $\nu = c/\lambda \Rightarrow c = \lambda\nu$. Equation [1.388] can be written in the form:

$$\nu^2 \frac{\partial^2 \Psi}{\partial q^2} = \frac{1}{\lambda^2} \frac{\partial^2 \Psi}{\partial t^2} \quad [1.389]$$

(2) *Proof*

The solution to the propagation equation [1.389] has the form:

$$\Psi(q,t) = \Phi(q) \times \exp(i\omega t) \quad [1.390]$$

According to the De Broglie relation, $\lambda = h/p$. Using this relation, the wave equation [1.389] is written as:

$$\nu^2 \frac{\partial^2 \Psi}{\partial q^2} = \frac{p^2}{h^2} \frac{\partial^2 \Psi}{\partial t^2} \quad [1.391]$$

Moreover, using [1.390], the second-order derivatives of the wave function with respect to the generalized coordinate q and with respect to time are:

$$\begin{cases} \dfrac{\partial^2 \Psi(q,t)}{\partial q^2} = \dfrac{\partial^2 \Phi(q,t)}{\partial q^2} \times \exp(i\omega t) \\ \dfrac{\partial^2 \Psi(q,t)}{\partial t^2} = -\omega^2 \Phi(q) \times \exp(i\omega t) \end{cases} \quad [1.392]$$

Using [1.392], after simplification equation [1.391] becomes:

$$\nu^2 \frac{\partial^2 \Phi(q,t)}{\partial q^2} = -\frac{p^2}{h^2} \omega^2 \Phi(q) \quad [1.393]$$

Since $\omega = 2\pi\nu$, we finally find:

$$\frac{\partial^2 \Phi(q)}{\partial q^2} = -4\pi^2 \frac{p^2}{h^2} \Phi(q) \quad [1.394]$$

This equation is valid for a free particle.

(3) *Expression of energy*

We generalize equation [1.394] to the case of a particle subjected to a potential. For this, we consider the particular case of a conservative system subjected to potential $V(q)$. Energy E is given by the relation:

$$E = \frac{p^2}{2m} + V(q) \qquad [1.395]$$

This expression gives the total energy of a non-relativistic particle (which already gives the idea that the Schrödinger equation is a non-relativistic equation).

(4) *Schrödinger's equation for stationary states*

Expression [1.395] can be used to deduce the linear momentum p of the particle:

$$p^2 = 2m[E - V(q)] \qquad [1.396]$$

Inserting [1.396] into [1.394], and knowing that $h^2/4\pi = \hbar^2$, we find:

$$-\hbar^2 \frac{\partial^2 \Phi(q)}{\partial q^2} = 2m[E - V(q)]\Phi(q)$$

or, after arrangement:

$$-\frac{\hbar^2}{2m}\frac{\partial^2 \Phi(q)}{\partial q^2} = [E - V(q)]\Phi(q) \qquad [1.397]$$

This is the stationary Schrödinger equation.

(5) *Generalization to three dimensions*

Equation [1.397] can be written in three dimensions (x, y, z) as follows:

$$\begin{cases} -\dfrac{\hbar^2}{2m}\dfrac{\partial^2 \Phi(x,y,z)}{\partial x^2} = [E_x - V(x)]\Phi(x,y,z) \\ -\dfrac{\hbar^2}{2m}\dfrac{\partial^2 \Phi(x,y,z)}{\partial y^2} = [E_y - V(y)]\Phi(x,y,z) \\ -\dfrac{\hbar^2}{2m}\dfrac{\partial^2 \Phi(x,y,z)}{\partial z^2} = [E_z - V(z)]\Phi(x,y,z) \end{cases} \qquad [1.398]$$

Summing equations [1.398], we find:

$$-\frac{\hbar^2}{2m}\left[\frac{\partial^2}{\partial x^2}+\frac{\partial^2}{\partial y^2}+\frac{\partial^2}{\partial z^2}\right]\Phi(x,y,z)=\left[(E_x+E_y+E_z)-\{V(x)+V(y)+V(z)\}\right]\Phi(x,y,z)$$

Making use of the Laplacian, we have:

$$-\frac{\hbar^2}{2m}\nabla^2\Phi(\vec{r})=\left[E-V(\vec{r})\right]\Phi(\vec{r})$$

which can be finally written as:

$$\left[-\frac{\hbar^2}{2m}\nabla^2+V(\vec{r})\right]\Phi(\vec{r})=E\Phi(\vec{r}) \qquad [1.399]$$

This is the stationary Schrödinger equation [1.28].

2

Hermitian Operator, Dirac's Notations

General objective

The general objective is to know the properties of Hermitian operators and the usefulness of Dirac's notations.

Specific objectives

On completing this chapter, the reader should be able to:

– define the space of square-summable wave functions;

– know the properties of the scalar product of two functions;

– define a discrete orthonormal basis;

– define the Kronecker symbol;

– define the components of a wave function;

– define the norm of a wave function;

– know the orthonormalization relation;

– write the expansion of a wave function;

– know the closing relation;

– define the space of states;

– know Dirac's notations;

For color versions of the figures in this book, see www.iste.co.uk/sakho/quantum2.zip

- write the expansion of a state vector;
- distinguish between a ket vector and a bra vector;
- distinguish between a linear operator and a linear functional;
- define the components of a ket and a bra;
- define a matrix element;
- define the projection operator on a ket and on a sub-space;
- define a self-adjoint operator;
- define a Hermitian operator;
- give examples of Hermitian operators;
- know the rules of Hermitian conjugation;
- define a function of operators;
- know the commutation rules;
- define the Poisson brackets;
- know the properties of commutators;
- define the trace of an operator;
- define a unitary operator;
- define the density operator;
- define the evolution operator;
- define an observable;
- know the properties of observables associated with spin;
- know the properties of Pauli matrices;
- know the properties of an orbital angular momentum operator.

Prerequisites

- Wave function.
- Properties of Cartesian space.
- Vector space.

2.1. Orthonormal bases in the space of square-summable wave functions

2.1.1. *Subspace of square-summable wave functions*

As already noted in Chapter 1, the wave function $\Psi(\vec{r})$ describing the physical state of a system satisfies the normalization condition (see [4.51], Chapter 4, Volume 1). The set of square-summable wave functions belongs to Hilbert space L^2 [COH 77, MAR 00, HLA 00, NEU 18]. Since the dimension of Hilbert space is too large, a subspace of L^2 denoted F is considered. This is constituted of square-summable wave functions that are defined everywhere, continuous and indefinitely differentiable. As the set F has the structure of a vector space, every wave function $\Psi(\vec{r}) \in F$ satisfies the superposition principle [1.1].

In addition, if $\Phi(\vec{r}) \in F$ and $\psi(\vec{r}) \in F$, the scalar product of $\Phi(\vec{r})$ and $\psi(\vec{r})$ considered in this order is a complex number and denoted as (ψ,Φ) and given by the relation [1.2] (this will be widely used in this chapter). The properties of the scalar product are expressed by the relations [1.3].

2.1.2. *Definition of discrete orthonormal bases*

Let $\{u_i(\vec{r})\} \in F$ be a discrete set, where i is a discrete index: $i = 1, 2, 3\ldots$.

$\{u_i(\vec{r})\}$ is an orthonormal set if:

$$\begin{cases} (u_i, u_j) = \delta_{ij} \\ (u_i, u_j) = \int d^3 r\, u_i^*(\vec{r}) u_j(\vec{r}) = \delta_{ij} \end{cases} \qquad [2.1]$$

where:

$$\delta_{ij} = \begin{cases} 1 \ if \ i = j \\ 0 \ if \ i \neq j \end{cases}$$

δ_{ij} is defined as the *Kronecker symbol*. The name of this symbol honors the German mathematician and logician **Léopold Kronecker** (1823–1891).

The set $\{u_i(\vec{r})\}$ constitutes a basis, if any wave function $\Psi \in F$ is uniquely expanded on the basis vectors $u_i(\vec{r})$ such that:

$$\begin{cases} \Psi(\vec{r}) = \sum_i c_i u_i(\vec{r}) \\ \Psi(\vec{r},t) = \sum_i c_i(t) u_i(\vec{r}) \end{cases} \qquad [2.2]$$

2.1.3. *Component and norm of a wave function*

Let us consider the expansion [2.2] and then express the scalar product (u_j,Ψ) using [1.2]. We obtain:

$$(u_j,\Psi) = (u_j, \sum_i c_i u_i) = \sum_i c_i (u_j,u_i) = \sum_i c_i \delta_{ji}$$

or:

$$c_j = (u_j,\Psi) \Rightarrow c_i = (u_i,\Psi) \qquad [2.3]$$

Result [2.3] shows that component c_i of the wave function $\Psi(\vec{r})$ on the discrete basis $\{u_i(\vec{r})\}$ is equal to the scalar product of $u_i(\vec{r})$ and $\Psi(\vec{r})$.

Let us now consider two functions Ψ and Φ of F and then express the scalar product (Ψ, Φ) as a function of their respective components b_i and c_j. We obtain:

$$\begin{cases} \Psi = \sum_i b_i u_i \\ \Phi = \sum_j c_j u_j \end{cases} \Rightarrow (\Psi,\Phi) = \left(\sum_i b_i u_i, \sum_j c_j u_j \right) = \sum_{i,j} b_i^* c_j (u_i,u_j)$$

Taking [2.1] into account, we have:

$$(\Psi,\Phi) = \sum_{i,j} b_i^* c_j \delta_{ij} \Rightarrow (\Psi,\Phi) = \sum_i b_i^* c_i \qquad [2.4]$$

In particular, if $\Phi = \Psi$, the squared norm of the wave function is (Ψ,Ψ), with:

$$(\Psi,\Psi) = \sum_i |c_i|^2 \qquad [2.5]$$

APPLICATION 2.1.–

Calculate the norm of the wave function below and draw a conclusion.

$$\Psi = \frac{1}{\sqrt{2}}u_1 - \frac{i}{2}u_2 \qquad [2.6]$$

Solution. We use [2.5]:

$$(\Psi,\Psi) = \sum_i |c_i|^2 = |c_1|^2 + |c_2|^2 = \left|\frac{1}{\sqrt{2}}\right|^2 + \left|-\frac{i}{2}\right|^2 \qquad [2.7]$$

$$(\Psi,\Psi) = \frac{1}{2} + \frac{1}{4} = \frac{3}{4}$$

Hence:

$$\sqrt{(\Psi,\Psi)} = \sqrt{\frac{3}{4}} = \frac{\sqrt{3}}{2} \qquad [2.8]$$

CONCLUSION.– The wave function is not normed to unity.

2.1.4. *Closing relation*

The orthonormalization relation [2.1] reflects the fact that the basis vectors $u_i(\vec{r})$ are orthogonal and normed to unity. In what follows, a relation will be established expressing that the set $\{u_i(\vec{r})\}$ is a basis in the space of square-summable wave functions. Taking [2.2] and [2.3] into consideration, we have:

$$\begin{cases} \Psi(\vec{r}) = \sum\limits_i c_i u_i(\vec{r}) \\ c_i = (u_i, \Psi) \end{cases} \Rightarrow \Psi = \sum_i (u_i, \Psi) u_i(\vec{r})$$

Using [2.3] in terms of variable $\vec{r}'$, we have:

$$c_i = \left(u_i(\vec{r}'), \Psi(\vec{r}')\right) = \int d^3r' u_i^*(\vec{r}')\Psi(\vec{r}') \qquad [2.9]$$

Hence:

$$\Psi(\vec{r}) = \sum_i \int d^3r' u_i^*(\vec{r}')\Psi(\vec{r}')u_i(\vec{r})$$

This means:

$$\Psi(\vec{r}) = \int d^3r' \left[\sum_i u_i^*(\vec{r}')\, u_i(\vec{r}) \right] \Psi(\vec{r}') \qquad [2.10]$$

By definition, *Dirac's delta function* (δ) has the property that, for any wave function $\Psi(\vec{r})$, satisfies the equation:

$$\Psi(\vec{r}) = \int d^3r' \Psi(\vec{r}')\delta(\vec{r} - \vec{r}') \qquad [2.11]$$

The comparison of expansions [2.10] and [2.11] reveals that the function u_i must satisfy the relation:

$$\sum_i u_i^*(\vec{r}')\, u_i(\vec{r}) = \delta(\vec{r} - \vec{r}') \qquad [2.12]$$

The *closing relation* [2.12] expresses the fact that the set $\{u_i(\vec{r})\}$ constitutes a basis in the space of square-summable wave functions.

2.2. Space of states, Dirac's notations

2.2.1. *Definition*

The previous section focused on the study of the properties of wave functions $\Psi(\vec{r}) \in F$. In quantum mechanics, the state of a physical system is described by a state vector denoted $|\Psi\rangle$ or $|\Psi(t)\rangle$ belonging to the space E_r of states. The passage from the *space of square-summable wave functions* to the *space of states* can be formally expressed as follows:

$$\begin{aligned} \Psi(\vec{r}) \in F \quad &\rightarrow \quad |\Psi\rangle \in E_r \\ \Psi(\vec{r},t) \in F \quad &\rightarrow \quad |\Psi(t)\rangle \in E_r \end{aligned} \qquad [2.13]$$

Let us note that in the writing of the state vector, the dependence with respect to variable $\vec{r}$ is no longer present. Only the temporal variable t is highlighted in the symbol of the state vector. The wave function $\Psi(\vec{r})$ is then interpreted as the set of components of the state vector on a particular basis denoted $\{|\vec{r}\rangle\}$ in which $\vec{r}$ plays the role of a continuous index, the components of $\vec{r}$ varying between $-\infty$ and $+\infty$. The relation of the passage from the state vector $|\Psi\rangle$ to the wave function $\Psi(\vec{r})$ will be revisited in section 2.2.2.

2.2.2. *Ket vector, bra vector*

The symbol $|\rangle$ involved in writing the state vector is known as *ket* and the state vector $|\Psi\rangle$ or $|\Psi(t)\rangle$ is known as a *ket vector* or simply ket. Therefore, any square-summable wave function $\Psi(\vec{r})$ is associated with a ket vector denoted $|\Psi\rangle$ belonging to the space of states E_r. By definition, a bra $\langle\Phi|$ is a *linear functional* that establishes a correspondence between any ket $|\Psi\rangle$ and a complex number λ such that:

$$(\langle\Phi|)|\Psi\rangle = \langle\Phi|\Psi\rangle = \lambda \qquad [2.14]$$

It can be proven that the set of linear functional defines a vector space denoted by E^*_r known as the *dual* of E_r.

Paul Adrien Maurice Dirac was a British physicist. In 1927, he published the results concerning the statistical distribution of the half-integer spin particles known as fermions in cooperation with the Italian physicist **Enrico Fermi** (1901–1954). This led to the elaboration of the Fermi-Dirac distribution or statistics, which is commonly employed in the physics of semiconductors. In 1928, Dirac established the relativistic wave equation in order to unify quantum mechanics and the theory of special relativity. The development of this relativistic theory of quantum mechanics made it possible to predict in 1931 the existence of a positron (positive electron), which was discovered in 1932 by the American physicist **Carl David Anderson** (1905–1991). In 1930, Dirac introduced linear operator algebra as a generalization of Heisenberg's and Schrödinger's theories. He also introduced the notions of ket and bra known as Dirac's notations, greatly simplifying the mathematical formalism of quantum mechanics. In 1933, he was awarded the Nobel Prize for physics, which he shared with Schrödinger for their important contributions to quantum mechanics.

Box 2.1. *Dirac (1902–1984)*

The symbol $\langle | \rangle$ is known as bracket. This explains the origin of the names ket for the symbol $| \rangle$ and bra for the symbol $\langle |$. The ket $| \rangle$ and bra $\langle |$ notations are known as Dirac's notations to honor Paul Dirac who introduced them to quantum mechanics.

2.2.3. *Properties of the scalar product*

The scalar product has been defined for two square-summable wave functions [1.2] and its properties have also been established [1.3]. They are established here once again in the space of states. Using Dirac's notations, the scalar product of ket $|\Psi\rangle$ and ket $|\Phi\rangle$ is defined by the relation:

$$(|\Phi\rangle, |\Psi\rangle) = \langle\Phi|\Psi\rangle \qquad [2.15]$$

This scalar product verifies the following properties:

$$\begin{cases} \langle\Phi|\Psi\rangle^* = \langle\Psi|\Phi\rangle \\ \langle\Psi|\lambda_1\varphi_1 + \lambda_2\varphi_2\rangle = \lambda_1\langle\varphi_1|\Psi\rangle + \lambda_2\langle\varphi_2|\Psi\rangle \\ \langle\lambda_1\varphi_1 + \lambda_2\varphi_2|\Psi\rangle = \lambda_1^*\langle\varphi_1|\Psi\rangle + \lambda_2^*\langle\varphi_2|\Psi\rangle \end{cases} \qquad [2.16]$$

If $\langle\Phi|\Psi\rangle = 0$, then ket $|\Phi\rangle$ and ket $|\Psi\rangle$ are orthogonal.

If $|\Phi\rangle = |\Psi\rangle$, then the squared norm of the ket is equal to $\langle\Psi|\Psi\rangle$.

If the ket $|\Psi\rangle$ is normed to unity, then $\langle\Psi|\Psi\rangle = 1$.

2.2.4. *Discrete orthonormal bases, ket component*

Discrete orthonormal bases have been previously defined [2.3] for the space of square-summable wave functions. They are redefined here in the space of states and their properties are studied using Dirac notations.

Let us consider a discrete set $\{|u_i\rangle\} \in E_r$, where i is a discrete index: $i = 1, 2, 3, \ldots$.

The set $\{|u_i\rangle\}$ is an orthonormal set if:

$$\langle u_i|u_j\rangle = \delta_{ij} \qquad [2.17]$$

δ_{ij} is the previously defined [2.3] Kronecker symbol.

The set $\{|u_i\rangle\}$ constitutes a basis in the space of states if any ket $|\Psi\rangle$ in the space of states E_r is uniquely expanded on $\{|u_i\rangle\}$, hence:

$$|\Psi\rangle = \sum_i c_i |u_i\rangle \quad [2.18]$$

Let us determine the scalar product ($|u_j\rangle$, $|\Psi\rangle$). Using [2.18], we get:

$$\left(|u_j\rangle, |\Psi\rangle\right) = \left(|u_j\rangle, \sum_i c_i |u_i\rangle\right) = \sum_i c_i \langle u_j | u_i\rangle = \sum_i c_i \delta_{ji}$$

which means:

$$(|u_j\rangle, |\Psi\rangle) = \langle u_j|\Psi\rangle = c_j \quad [2.19]$$

Therefore, the component c_i of the state vector $|\Psi\rangle$ on the basis $\{|u_i\rangle\}$ is equal to the scalar product of $|\Psi\rangle$ and $|u_i\rangle$, hence:

$$c_i = \langle u_i|\Psi\rangle \quad [2.20]$$

Moreover, relation [2.17] expresses the fact that the set $\{|u_i\rangle\}$ is orthonormal. We will establish a further relation that expresses that this set is a basis in the space of states. For this purpose, we substitute c_i by its expression [2.20] in the expansion [2.18] of the state vector. We obtain:

$$|\Psi\rangle = \sum_i \langle u_i|\Psi\rangle |u_i\rangle = \sum_i |u_i\rangle\langle u_i|\Psi\rangle = \left(\sum_i |u_i\rangle\langle u_i|\right)|\Psi\rangle$$

Knowing that $|\Psi\rangle$ is arbitrary, the *closing relation* satisfied by the set $\{|u_i\rangle\}$ can then be written as:

$$\sum_i |u_i\rangle\langle u_i| = \mathbb{1} \quad [2.21]$$

Relation [2.21] is the equivalent of the closing relation [2.12] satisfied by the set $\{u_i(\vec{r})\}$ in the space of the square-summable wave functions.

2.3. Hermitian operators

2.3.1. *Linear operator, matrix element*

Linear operators are similar to those defined in the space of square-summable functions. They are redefined here using Dirac notations.

According to a Dirac notation system, a linear operator A is a mathematical being that establishes a correspondence between any ket $|\Psi\rangle$ of the space of states E_r and another ket $|\Psi'\rangle$ belonging to the same space, the correspondence being linear. Hence:

$$\begin{cases} A|\Psi\rangle = |\Psi'\rangle \\ A(\lambda_1|\psi_1\rangle + \lambda_2|\psi_2\rangle) = \lambda_1 A|\psi_1\rangle + \lambda_2 A|\psi_2\rangle \end{cases} \quad [2.22]$$

Properties [2.22] reflect the action of operator A on the kets. While operator A acts on the left of a ket, it always acts on the right of a bra. The action of operator A on the bras can then be written as follows:

$$(\lambda_1\langle\psi_1| + \lambda_2\langle\psi_2|)A = \lambda_1\langle\psi_1|A + \lambda_2\langle\psi_2|A \quad [2.23]$$

Expressing the scalar product of ket $A|\Psi\rangle$ and ket $|\Phi\rangle$, we obtain:

$$(|\Phi\rangle, \mathrm{A}|\Psi\rangle) = \langle\Phi|(\mathrm{A}|\Psi\rangle) = (\langle\Phi|\mathrm{A})|\Psi\rangle = \langle\Phi|\mathrm{A}|\Psi\rangle \quad [2.24]$$

By definition, a *matrix element* of A between kets $|\Phi\rangle$ and $|\Psi\rangle$ is the complex number denoted by [COH 77]:

$$\langle\Phi|A|\Psi\rangle \quad [2.25]$$

2.3.2. *Projection operator on a ket and projection operator on a subspace*

In the previous section, we have defined the linear operator using Dirac notations. Let us give a simple example of the linear operator known as the projection operator, which will be defined in the space of states. An interpretation of the action of such an operator on the kets will then be provided by an analogy with the Cartesian space.

Let us consider the quantity $P_\psi = |\psi\rangle\langle\psi|$ and then express its action on the ket $|\Psi\rangle$. We obtain:

$$P_\psi|\Psi\rangle = |\psi\rangle\langle\psi|\Psi\rangle = \langle\psi|\Psi\rangle|\psi\rangle = \lambda|\psi\rangle, \text{ with } \lambda = \langle\psi|\Psi\rangle$$

Hence

$$P_\psi|\Psi\rangle = \lambda|\psi\rangle = |\Psi'\rangle \quad [2.26]$$

Result [2.26] shows that the quantity P_ψ establishes a correspondence between any ket $|\Psi\rangle$ and another ket $|\Psi'\rangle$: $P_\psi = |\psi\rangle\langle\psi|$ is therefore an operator. Let us prove that it is linear.

Let us consider the action of P_ψ on the ket $\lambda_1|\varphi_1\rangle + \lambda_2|\varphi_2\rangle$. We obtain:

$$P_\psi\,(\lambda_1|\varphi_1\rangle + \lambda_2|\varphi_2\rangle) = |\psi\rangle\langle\psi|(\lambda_1|\varphi_1\rangle + \lambda_2|\varphi_2\rangle)$$

Hence:

$$|\psi\rangle\langle\psi|(\lambda_1|\varphi_1\rangle + \lambda_2|\varphi_2\rangle) = \lambda_1|\psi\rangle\langle\psi|\varphi_1\rangle + \lambda_2|\psi\rangle\langle\psi|\varphi_2\rangle$$

And finally:

$$P_\psi(\lambda_1|\varphi_1\rangle + \lambda_2|\varphi_2\rangle) = \lambda_1 P_\psi|\varphi_1\rangle + \lambda_2 P_\psi|\varphi_2\rangle \qquad [2.27]$$

Result [2.27] actually expresses the fact that operator P_ψ is linear.

Before providing a "geometrical" interpretation of relation [2.26], let us make sure that P_ψ is actually a projection operator. If P_ψ is a projection operator, then the following property is verified: $(P_\psi)^2 = P_\psi$.

We then consider a normed ket $|\psi\rangle$ that means: $\langle\psi|\psi\rangle = 1$. The squared operator P_ψ is written as:

$$(P_\psi)^2 = P_\psi \times P_\psi = (|\psi\rangle\langle\psi|) \times (|\psi\rangle\langle\psi|)$$

Hence:

$$(P_\psi)^2 = |\psi\rangle\langle\psi|\psi\rangle\langle\psi| = |\psi\rangle\langle\psi| = P_\psi$$

CONCLUSION.– From a "geometrical" point of view, P_ψ is the "*orthogonal projection operator*" on the ket $|\psi\rangle$ (see Figure 2.1).

As shown in Figure 2.1, the projection operator on the ket $|\varphi\rangle$ can also be defined. The projection operator on the two-dimensional subspace generated by the kets $|\psi\rangle$ and $|\varphi\rangle$ is then defined by the relation:

$$P_\psi + P_\varphi = |\psi\rangle\langle\psi| + |\varphi\rangle\langle\varphi| \qquad [2.28]$$

In order to generalize result [2.28], let us consider the subspace E_q generated by q basis vectors $\{|\varphi_i\rangle\}$, $i = 1, 2, 3,\ldots, q$ with $\langle\varphi_i|\varphi_j\rangle = \delta_{ij}$. Let P_q be the projection operator defined by the relation:

$$P_q = \sum_{i=1}^{q} |\varphi_i\rangle\langle\varphi_i| \qquad [2.29]$$

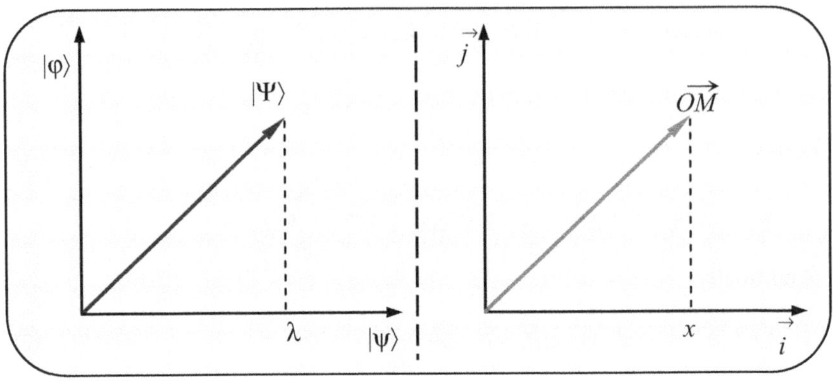

Figure 2.1. *"Geometrical" interpretation of the projection operator P_ψ on the ket $|\psi\rangle$*

By analogy to orthogonal projection in the Cartesian space:

$P_{(\vec{i})}\overrightarrow{OM} = (\vec{i}\cdot\overrightarrow{OM})\vec{i} = x\vec{i}$, we have: $P_\psi|\Psi\rangle = \langle\psi|\Psi\rangle|\psi\rangle = \lambda|\psi\rangle$

Let us verify that P_q is actually a projection operator. We obtain:

$$P_q^2 = \left(\sum_{i=1}^{q} |\varphi_i\rangle\langle\varphi_i|\right)\times\left(\sum_{j=1}^{q} |\varphi_j\rangle\langle\varphi_j|\right) = \sum_{i,j}^{q} |\varphi_i\rangle\langle\varphi_i|\varphi_j\rangle\langle\varphi_j| = \sum_{i,j}^{q} |\varphi_i\rangle\langle\varphi_j|\langle\varphi_i|\varphi_j\rangle$$

This finally leads to:

$$P_q^2 = \sum_{i,j}^{q} |\varphi_i\rangle\langle\varphi_j|\delta_{ij} = \sum_{i}^{q} |\varphi_i\rangle\langle\varphi_i| = P_q \qquad [2.30]$$

We then prove that P_q is a projection operator acting all over the subspace E_q. For this purpose, we consider the ket $|\Psi\rangle$ and we obtain:

$$P_q|\Psi\rangle = \left(\sum_{i=1}^{q} |\varphi_i\rangle\langle\varphi_i|\right)|\Psi\rangle = \sum_{i=1}^{q} |\varphi_i\rangle\langle\varphi_i|\Psi\rangle = \sum_{i=1}^{q} \langle\varphi_i|\Psi\rangle|\varphi_i\rangle$$

Hence:

$$P_q|\Psi\rangle = \sum_{i=1}^{q} \lambda_i|\varphi_i\rangle = \lambda_1|\varphi_1\rangle + \lambda_2|\varphi_2\rangle + \lambda_3|\varphi_3\rangle + ...\lambda_q|\varphi_q\rangle \qquad [2.31]$$

Result [2.31] actually proves that projection operator P_q acts on the subspace E_q subtended by the q basis vectors $\{|\varphi_i\rangle\}$.

2.3.3. *Self-adjoint operator, Hermitian conjugation*

Let A be a linear operator. The ket:

$$A|\Psi\rangle \in E_r \to (A|\Psi\rangle)^* \in E^*_r \qquad [2.32]$$

with:

$$(A|\Psi\rangle)^* = \langle\Psi|A^\dagger \qquad [2.33]$$

This is the definition of the adjoint operator $A^\dagger$ of A that uses Dirac's notations. The matrix element [2.24] then leads to the simple relation:

$$\langle\Phi|A|\Psi\rangle^* = \langle\Psi|A^\dagger|\Phi\rangle \qquad [2.34]$$

If A is linear, then $A^\dagger$ is also linear. If λ is a complex number, the following *Hermitian conjugations* using Dirac's notations are verified:

$$(\lambda A)^\dagger = \lambda^* A^\dagger;\ (\lambda^* A^\dagger)^\dagger = \lambda A \qquad [2.35]$$

$$(\langle\Phi|A\lambda\Psi\rangle)^* = \lambda^*\langle\Psi|A^\dagger|\Phi\rangle \qquad [2.36]$$

$$(\langle\lambda_1\varphi_1+\lambda_2\varphi_2|A)^* = A^\dagger|\lambda_1\varphi_1+\lambda_2\varphi_2\rangle = \lambda_1 A^\dagger|\varphi_1\rangle + \lambda_2 A^\dagger|\varphi_2\rangle \qquad [2.37]$$

$$(\lambda\langle\varphi|A|\Psi\rangle)^* = \lambda^*\langle\Psi|A^\dagger|\varphi\rangle \qquad [2.38]$$

As a general rule, in order to obtain the *Hermitian conjugate* of an arbitrary expression that involves kets, bras, operators and complex numbers, it is sufficient to write the vectors and operators in reverse order replacing:

– complex numbers by their complex conjugates;

– kets by their associated bras;

– bras by their associated kets;

– operators by their adjoints and the inverse.

The order in which the complex numbers and their conjugates are placed is not important.

If operator A is Hermitian, then $A = A^\dagger$. Relation [2.34] is then written as:

$$\langle\Phi|A|\Psi\rangle^* = \langle\Psi|A|\Phi\rangle \qquad [2.39]$$

APPLICATION **2.2.**–

Prove the following property:

$$(AB)^\dagger = B^\dagger A^\dagger \qquad [2.40]$$

Solution. Given the ket $|\Psi\rangle = (AB)|\Phi\rangle = A(B|\Phi\rangle)$, we put: $B|\Phi\rangle = |\psi\rangle$; $|\Psi\rangle = A|\psi\rangle$.

We then obtain:

$$\langle\Psi| = \langle\Phi|(AB)^\dagger = \langle\psi|A^\dagger = \langle\Phi|B^\dagger A^\dagger \Rightarrow (AB)^\dagger = B^\dagger A^\dagger$$

Therefore, when the *adjoint of the product of two operators* is set, the initial order of the operators is reversed.

2.3.4. *Operator functions*

An example of an operator function has already been provided in Chapter 1. It is the *evolution operator* denoted $U(t, t_0)$ defined by relation [1.34]. The objective is here to define operator functions in the general case and then express several of their properties.

Let A be a linear operator and $F(u)$ a function of the variable u that can be expanded in the form of an integer series such that:

$$F(u) = \sum_{n=0}^{\infty} f_n u^n \qquad [2.41]$$

By definition, the operator $F(A)$ of the linear operator A is defined by the expression:

$$F(A) = \sum_{n=0}^{\infty} f_n A^n \qquad [2.42]$$

If A is Hermitian, then $F(A)$ is Hermitian if coefficients f_n are real.

APPLICATION 2.3.–

Let A be an observable. Prove that the operator $F(A) = e^A$ is Hermitian.

Solution. Let us expand the operators F(A) and $F(A^\dagger)$ in integer series using relation [2.42]. We obtain:

$$e^A = \sum_{n=0}^{\infty} \frac{1}{n!} A^n = \mathbb{1} + \frac{1}{2!} A^2 + \frac{1}{3!} A^3 + \ldots \frac{1}{n!} A^n + \ldots \qquad [2.43]$$

$$e^{A\dagger} = \sum_{n=0}^{\infty} \frac{1}{n!} (A^\dagger)^n = \mathbb{1} + \frac{1}{2!} (A^\dagger)^2 + \frac{1}{3!} (A^\dagger)^3 + \ldots \frac{1}{n!} (A^\dagger)^n + \ldots$$

Since A is Hermitian, then:

$$e^{A\dagger} = \sum_{n=0}^{\infty} \frac{1}{n!} (A^\dagger)^n = \mathbb{1} + \frac{1}{2!} (A)^2 + \frac{1}{3!} (A)^3 + \ldots \frac{1}{n!} (A)^n + \ldots \qquad [2.44]$$

Comparing [2.43] and [2.44], it can be noted that F (A) = $F(A^\dagger)$: F (A) is actually Hermitian.

2.4. Commutator algebra

2.4.1. *Poisson brackets*

As already noted throughout Chapter 1, the operators representing fundamental observables are the position and linear momentum operators, which are used in building various other operators, such as the Hamiltonian, based on the correspondence principle stated in section 1.3.4.

Let us consider two arbitrary functions $F(q_k, p_k)$ and $G(q_k, p_k)$ defined in the space of phases characterized by the *canonical coordinates* q_k (*generalized coordinate*) and p_k (*generalized linear momentum*), i =1, 2, 3… *Poisson's bracket* is defined as the quantity denoted [F, G], which is given by the expression [CHP 78]:

$$[F,G]=\sum_k\left(\frac{\delta F}{\delta p_k}\frac{\delta G}{\delta q_k}-\frac{\delta G}{\delta p_k}\frac{\delta F}{\delta q_k}\right) \quad [2.45]$$

Let λ be a complex number. Poisson's bracket verifies the following properties:

$$[F, F] = [G, G] = 0;\ [F, \lambda] = [G, \lambda] = 0;\ [F, G] = -[G, F] \quad [2.46]$$

$$[F, G+K] = [F, G] + [F, K];\ [G+K, F] = [G, F] + [K, F] \quad [2.47]$$

$$[F, GK] = [F, G]\, K + G[F, K];\ [GK, F] = G[K, F] + [G, F]K \quad [2.48]$$

The properties below are also satisfied if F and G are operators and if the operator [F, G] represents their commutator (see further below).

APPLICATION 2.4.–

Prove the first of relations [2.47].

Solution. Using [2.45], we have:

$$[F,G+K]=\sum_k\left(\frac{\delta F}{\delta p_k}\frac{\delta(G+K)}{\delta q_k}-\frac{\delta(G+K)}{\delta p_k}\frac{\delta F}{\delta q_k}\right)$$

Hence:

$$[F,G+K]=\sum_k\left(\frac{\delta F}{\delta p_k}\frac{\delta G}{\delta q_k}-\frac{\delta G}{\delta p_k}\frac{\delta F}{\delta q_k}\right)+\sum_k\left(\frac{\delta F}{\delta p_k}\frac{\delta K}{\delta q_k}-\frac{\delta K}{\delta p_k}\frac{\delta F}{\delta q_k}\right)$$

Or:

$$[F, G+K] = [F, G] + [F, K]$$

APPLICATION 2.5.–

Prove that the product (AB) of two Hermitian operators A and B is not Hermitian unless the two operators commute.

Solution. Let A and B be two arbitrary operators. The commutator of A and B is the operator denoted $[A, B]$ defined by the relation:

$$[A, B] = AB - BA \tag{2.49}$$

If A and B commute, then:

$$[A, B] = 0 \Rightarrow AB = BA \tag{2.50}$$

If A and B are Hermitian, then according to [2.40] we have:

$$(AB)^\dagger = B^\dagger A^\dagger = BA \tag{2.51}$$

Knowing that A and B commute, then relation [2.52] can be written as:

$$(AB)^\dagger = (AB) \tag{2.52}$$

Therefore, the product (AB) of two Hermitian operators A and B is not Hermitian unless the two operators commute.

APPLICATION 2.6.–

The *quantum Poisson bracket* is defined as the operator $[F, G]$ given by the relation:

$$[F,G] = \frac{i}{\hbar}(FG - GF) \tag{2.53}$$

Prove that if F and G are Hermitian, then operator $[F, G]$ defined by [2.53] is Hermitian because of the introduction of factor i.

Solution. Using the property [2.40], we have:

$$[F,G]^\dagger = \left(\frac{i}{\hbar}(FG - GF)\right)^\dagger = -\frac{i}{\hbar}(FG - GF)^\dagger = -\frac{i}{\hbar}(G^\dagger F^\dagger - F^\dagger G^\dagger)$$

Hence:

$$[F,G]^\dagger = \left(\frac{i}{\hbar}(FG - GF)\right)^\dagger = -\frac{i}{\hbar}(FG - GF)^\dagger = \frac{i}{\hbar}(F^\dagger G^\dagger - G^\dagger F^\dagger) \tag{2.54}$$

If F and G are Hermitian, then:

$$[F,G]^{\dagger} = \frac{i}{\hbar}(FG - GF) = [F,G] \qquad [2.55]$$

The introduction of factor i in the definition relation [2.53] is a requirement for ensuring the Hermiticity of the quantum Poisson bracket.

2.4.2. *Commutation of operator functions*

As a general rule, any operator A commutes with its function $F(A)$. Moreover, if A and B commute, then:

$$[A, F(A)] = 0;\ [B, F(A)] = 0;\ [A, F(B)] = 0 \qquad [2.56]$$

Furthermore, as seen in Volume 1, Chapter 4, according to Heisenberg uncertainty relations, it is impossible to simultaneously measure the position x and the linear momentum $p = p_x$. In other terms, the associated operators X and P are anticommutative. To verify these assertions, let us first calculate commutator $[X, P]$.

We consider the continuous set $\{|x\rangle\}$ and express the number $\langle x|[X, P]|\psi\rangle$:

$$\langle x|[X,P]|\psi\rangle = \langle x|(XP - PX)|\psi\rangle = \langle x|XP|\psi\rangle - \langle x|PX|\psi\rangle$$

Hence:

$$\langle x|[X,P]|\psi\rangle = x\langle x|P|\psi\rangle - \langle x|Px|\psi\rangle = -i\hbar x\frac{d}{dx}\langle x|\psi\rangle + i\hbar x\frac{d}{dx}\langle x|\psi\rangle \qquad [2.57]$$

Expanding the left member of [2.57], we have:

$$\langle x|[X,P]|\psi\rangle = -i\hbar x\frac{d}{dx}\langle x|\psi\rangle + i\hbar x\frac{d}{dx}\langle x|\psi\rangle + i\hbar\langle x|\psi\rangle$$

After simplification and arrangement, we find:

$$\langle x|[X,P]|\psi\rangle = \langle x|i\hbar|\psi\rangle \qquad [2.58]$$

Which finally leads to:

$$[X,P] = i\hbar\mathbb{1} \qquad [2.59]$$

APPLICATION 2.7.–

Calculate the commutators $[X, P^2]$, $[X, P^3]$ and $[X, P^n]$, $n > 0$.

Solution.

– *Commutator* $[X, P^2]$

Using the first property [2.48], we obtain:

$$[X, P^2] = [X, PP] = [X, P]P + [X, P]P$$

Using [2.59] (for the sake of simplification, the identity operator is omitted), we find:

$$[X, P^2] = 2i\hbar P \qquad [2.60]$$

– *Commutator* $[X, P^3]$

Similarly, we obtain:

$$[X, P^3] = [X, P^2]P + P^2[X, P]$$

Using results [2.59] and [2.60], we find:

$$[X, P^3] = 3i\hbar P \qquad [2.61]$$

– *Commutator* $[X, P^n]$

Let us consider the commutator:

$$[X, P^{p+1}] = [X, PP^p] = [X, P]P^p + P[X, P^p]$$

According to [2.60] and [2.61], we obtain by recurrence:

$$[X, P^p] = pi\hbar P^{p-1}$$

This involves:

$$[X, P^{p+1}] = i\hbar P^p + pi\hbar PP^{p-1} = i\hbar(p+1)P^p$$

This relation is valid for any integer p in particular for $p = n - 1$. Hence:

$$[X, P^n] = i\hbar nP^{n-1} \qquad [2.62]$$

We now consider the function $F(P)$ of operator P defined by relation [2.42], in which the operator A is replaced by P. Then we express the commutator $[X, F(P)]$. We obtain:

$$[X,F(P)]=\sum_n f_n[X,P^n]=i\hbar\sum_n f_n\, n\, P^{n-1} \qquad [2.63]$$

The last term of equation [2.63] features the derivative of the operator function $F(P)$ with respect to P, which is:

$$F'(P)=\frac{dF(P)}{dP}=\frac{d}{dP}\sum_n f_n P^n=\sum_n f_n\, n\, P^{n-1}$$

Therefore, we finally get:

$$[X,F(P)]=i\hbar F'(P) \qquad [2.64]$$

Let us also express the commutator $[P, G(X)]$. In this commutator, $G(X)$ is a function of the position operator X defined by relation [2.42] in which A must be replaced by X. We obtain:

$$[P,G(X)]=\sum_n g_n[P,X^n] \qquad [2.65]$$

Or:

$$[P,X^n]=[P,XX^{n-1}]=[P,X]X^{n-1}+X[P,X^{n-1}] \qquad [2.66]$$

Using the third property [2.46], we obtain $[P,X]=-i\hbar$ according to [2.59]. Relation [2.66] is finally written as:

$$[P,X^n]=-i\hbar X^{n-1}-i\hbar X(n-1)X^{n-2}$$

Or after arrangement:

$$[P,X^n]=-i\hbar nX^{n-1} \qquad [2.67]$$

Inserting [2.67] in [2.65], we find:

$$[P,G(X)]=-i\hbar\sum_n g_n nX^{n-1}$$

The derivative of function $G(X)$ with respect to X is written as:

$$G'(X) = \frac{dG(X)}{dX} = \frac{d}{dX}\sum_n g_n X^n = \sum_n g_n \, n \, X^{n-1}$$

Hence:

$$[P, G(X)] = -i\hbar G'(X) \qquad [2.68]$$

Results [2.64] and [2.68] feature the derivatives of the operator functions $F(P)$ and $G(X)$ with respect to P and X, respectively. We now consider the case when F and G are time dependent. The rules of differentiation with respect to time t are identical to the rules of differentiation commonly used for classical quantities. Hence:

$$\frac{d(F+G)}{dt} = \frac{dF}{dt} + \frac{dG}{dt} \qquad [2.69]$$

$$\frac{d(FG)}{dt} = \frac{dF}{dt}G + F\frac{dG}{dt} \qquad [2.70]$$

The order of operators in relation [2.70] should be respected. If F and G commute, then this order has no importance.

APPLICATION 2.8.–

Let A and B be two time-dependent operators that do not commute. We consider the operator function $F(A, B, t) = e^{At}e^{Bt}$. Under what condition we can write:

$$\frac{dF}{dt} = (A+B)F \qquad [2.71]$$

Solution. Let us differentiate the function $F(A, B, t) = e^{At}e^{Bt}$ with respect to time. We get:

$$\frac{dF}{dt} = \frac{de^{At}}{dt}e^{Bt} + e^{At}\frac{de^{Bt}}{dt} \qquad [2.72]$$

Using [2.43], we have:

$$\frac{de^{Ct}}{dt} = \frac{d}{dt}\sum_{n=0}^{\infty}\frac{(Ct)^n}{n!} = \sum_{n=0}^{\infty}\frac{nC^n t^{n-1}}{n!}$$

Hence:

$$\frac{de^{Ct}}{dt} = C\sum_{n=1}^{\infty}\frac{(Ct)^{n-1}}{(n-1)!} = C\sum_{p=0}^{\infty}\frac{(Ct)^{p}}{p!} = Ce^{Ct} \quad [2.73]$$

Using [2.73], relation [2.72] can be written as:

$$\frac{dF}{dt} = Ae^{At}e^{Bt} + e^{At}Be^{Bt} \quad [2.74]$$

Relation [2.74] cannot be written in the form [2.71] unless A and B commute. They commute with their functions, hence $e^{At}Be^{Bt} = B\, e^{AtBt}$. Factoring the right member of [2.74] by e^{AtBt}, we actually obtain [2.71].

Siméon Denis Poisson was a French mathematician, physicist and geometrician. His contributions to physics essentially relate to electricity and magnetism. Relying on the **Laplace** (1749–1827) equation, **Poisson** published, in 1813, the differential equation satisfied by any electrostatic potential. This Poisson equation expresses that the Laplacian of the potential in a point in space depends only on the volume charge density in this point. Based on the notion of vector field flux, **Gauss** (1777–1855) formulated, in 1840, the local Poisson equation at macroscopic scale (Gauss theorem). In quantum mechanics, Poisson brackets are named in his honor.

Box 2.2. *Poisson (1781–1840)*

2.4.3. *Trace of an operator*

By definition, the *trace of an operator A,* denoted *TrA,* is the sum of its diagonal matrix elements. In the discrete orthonormal basis $\{|u_i\rangle\}$, the trace of operator A can be written as:

$$TrA = \sum_i \langle u_i|A|u_i\rangle \quad [2.75]$$

We prove that the trace of an operator is an invariant, meaning that it does not depend on the chosen basis. Then we consider another discrete basis $\{|u_k\rangle\}$ in the space of states. Let U be a matrix. A change of basis involves the passage from the set $\{|u_i\rangle\}$ to the set $\{|v_k\rangle\}$ by means of the transformation:

$$|u_i\rangle = \sum_k U_{ik}|v_k\rangle \quad [2.76]$$

Vectors $\{|v_k\rangle\}$ form a discrete orthonormal basis in the space of states, if matrix U is unitary, therefore:

$$\sum_k U_{ik}^* U_{jk} = \delta_{ij} \qquad [2.77]$$

Then we have:

$$Tr(A) = \sum_i \langle u_i | A | u_i \rangle = \sum_i \left(\sum_k U_{ik}^* \langle v_k | \right) A \left(\sum_l U_{il} | v_l \rangle \right)$$

Hence:

$$Tr(A) = \sum_{kl} \left(\sum_i U_{ik}^* U_{il} \right) \langle v_k | A | v_l \rangle = \sum_{kl} \langle v_k | A | v_l \rangle \delta_{kl}$$

It finally leads to:

$$Tr(A) = \sum_i \langle u_i | A | u_i \rangle = \sum_k \langle v_k | A | v_k \rangle \qquad [2.78]$$

Result [2.78] actually expresses the fact that the trace of an operator is an invariant.

2.5. Exercises

2.5.1. *Exercise 1 – Properties of commutators*

Let A, B and C be three operators.

(1) Prove the following properties:

a) $[A, B] = -[B, A]$; b) $[A, (B + C)] = [A, B] + [A, C]$

c) $[A, BC] = [A, B]C + B[A, C]$; d) $[A, B]^\dagger = [B^\dagger, A^\dagger]$

(2) Fill in the following equations (λ and μ are complex numbers):

a) $(A^\dagger)^\dagger = \ldots\ldots$; b) $(\lambda A)^\dagger = \ldots\ldots\ldots$; c) $(\lambda A^\dagger + \mu^* B)^\dagger = \ldots\ldots\ldots$;

d) $(AB^\dagger)^\dagger = \ldots\ldots\ldots$

(3) Prove the following relations [*for* $\Psi(x) \to 0$ *when* $x \to \infty$]:

a) $\left(\frac{d}{dx}\right)^{\dagger} = -\frac{d}{dx}$; b) $\left(\frac{d^2}{dx^2}\right)^{\dagger} = \frac{d^2}{dx^2}$; c) $P^{\dagger} = P$ (*linear momentum operator*)

2.5.2. *Exercise 2 – Trace of an operator*

Let A be a linear operator. We consider the case when A is an observable. Let $\left|u_n^i\right\rangle$ be a ket of A verifying the equation $A\left|u_n^i\right\rangle = a_n\left|u_n^i\right\rangle$, $i = 1, 2, 3, \ldots, g_n$, where g_n is the degree of degeneracy.

(1) Prove the relation:

$$Tr\,A = \sum_n g_n a_n$$

(2) Prove the following properties:

a) Tr AB = Tr BA; b) Tr ABC = Tr BCA = Tr CAB

2.5.3. *Exercise 3 – Function of operators*

Let A and B be two operators that commute. We consider the functions of operators $F(A)$ and $F(B)$ defined by:

$$F(A) = \sum_n f_n A^n \; ; \; F(B) = \sum_n g_n B^n$$

Let A^n be the operator obtained by n successive applications of operator A on itself and let $|\Phi\rangle$ be a ket of A such that $A|\Phi\rangle = a|\Phi\rangle$.

(1) Express the action of operator $F(A)$ on the ket $|\Phi\rangle$.

(2) Given the matrix:

$$\sigma_z = \begin{pmatrix} 1 & 0 \\ 0 & -1 \end{pmatrix}$$

Find e^{σ_z}.

(3) Find the commutators $[A, F(A)]$, $[A, F(B)]$ and $[F(A), F(B)]$.

2.5.4. Exercise 4 – Infinitesimal unitary operator

Let us consider *an infinitesimal unitary operator* $U(\varepsilon)$ defined by the relation:

$$U(\varepsilon) = (\mathbb{1} + A)^{\varepsilon}$$

In this relation, ε is an infinitesimal quantity.

(1) Prove that the product of two unitary operators is also unitary.

(2) Expand $U(\varepsilon)$ to the first order approximation.

(3) Express in this approximation, the products $U^{\dagger}(\varepsilon)U(\varepsilon)$ and $U(\varepsilon)U^{\dagger}(\varepsilon)$ as a function of ε, A and $A^{\dagger}$.

(4) Is the operator A Hermitian?

(5) Prove that there is a Hermitian operator B such that $B = F(A)$.

2.5.5. Exercise 5 – Properties of Pauli matrices

An orthonormal basis $\{|+\rangle\,;\,|-\rangle$ is chosen. In this basis, dimensionless operators σ_x, σ_y, and σ_z are represented by *Pauli matrices* and defined by:

$$\sigma_x = \begin{pmatrix} 0 & 1 \\ 1 & 0 \end{pmatrix};\ \sigma_y = \begin{pmatrix} 0 & -i \\ i & 0 \end{pmatrix};\ \sigma_z = \begin{pmatrix} 1 & 0 \\ 0 & -1 \end{pmatrix}$$

The spin angular momentum is given by the expression:

$$\vec{S} = \frac{\hbar}{2}\vec{\sigma}$$

(1) Prove that σ_x, σ_y, and σ_z are self-adjoint operators.

(2) Compare σ_x^2, σ_y^2 and σ_z^2.

(3) Express $Tr(\sigma_i)$, $i = x, y$ and z.

(4) Prove that $\sigma_x\sigma_y + \sigma_y\sigma_x = 0$ and that $\sigma_x\sigma_y - \sigma_y\sigma_x = 2i\sigma_z$.

(5) Express the commutators $[\sigma_x, \sigma_y]$, $[\sigma_y, \sigma_z]$ and $[\sigma_z, \sigma_x]$.

(6) Deduce from the above the commutators $[S_x, S_y]$, $[S_y, S_z]$ and $[S_z, S_x]$.

2.5.6. *Exercise 6 – Density operator*

Let us consider a discrete orthonormal basis $\{|u_n\rangle\}$ in the space of states of a particle. At instant t, a ket vector of this state can be written as:

$$|\Psi(t)\rangle = \sum_n c_n(t)|u_n\rangle$$

(1) Express the relation that must be satisfied by coefficients $c_n\,(t)$ so that the state vector $|\Psi(t)\rangle$ is normed (this is the assumption in what follows).

(2) The *density operator* is defined in the representation $\{|u_n\rangle\}$:

$$\rho\,(t) = |\Psi(t)\rangle\langle\Psi(t)|$$

Prove that $\rho\,(t)$ is Hermitian and idempotent (hence a projection operator).

(3) Find the trace of the density operator.

2.5.7. *Exercise 7 – Evolution operator*

Let us consider an observable A acting in the space of states with n dimensions. We designate by $|\Phi_k\rangle$, the kets of A such that $A|\Phi_k\rangle = a_k|\Phi_k\rangle$ ($k = 1, 2.. n$). Moreover, we consider a linear operator $U\,(k, m)$ defined by the relation:

$$U\,(k, m) = |\Phi_k\rangle\langle\Phi_m|$$

(1) Establish the relation between $U\,(k, m)$ and its adjoint.

(2) Find the commutator $[A, U\,(k, m)]$. Deduce from it $[A, U\,(k, k)]$.

(3) Prove that $Tr\ U\,(k, m) = \delta_{mk}$.

(4) Let B be an observable of matrix elements B_{km}. Prove that:

$$B = \sum_{k,m} B_{km}\, U(k,m)$$

(5) Prove the relation:

$$B_{km} = Tr\ \{BU^{\dagger}\,(k, m)\}$$

(6) Let us consider a normed ket $|\Psi(t)\rangle$ whose evolution in time is described by the equation:

$$|\Psi(t)\rangle = U(t, t_0)|\Psi(t_0)\rangle$$

In this equation, $U(t, t_0)$ is the evolution operator.

(6.1) Prove that $U(t, t_0)$ is a unitary operator.

(6.2) Using the Schrödinger equation, prove the relation:

$$i\hbar\frac{dU(t,t_0)}{dt} = HU(t,t_0)$$

In this relation, H is the Hamiltonian of a conservative system.

(6.3) Deduce from it the expression of the evolution operator $U(t, t_0)$.

2.5.8. *Exercise 8 – Orbital angular momentum operator*

We consider the orbital angular momentum operator defined by the relation:

$$\vec{l} = \vec{r} \wedge \vec{p}$$

The component p_q of the linear momentum operator is defined by the relation:

$$p_q = -i\hbar\frac{\partial}{\partial q}$$

(1) Express the components l_x, l_y and l_z of the angular momentum operator.

(2) Establish the expressions of the products of operators $l_x l_y$ and $l_y l_x$.

(3) Deduce from them the commutators $[l_i, l_j]$ $(i = x, y, z \neq j)$.

2.6. Solutions

2.6.1. *Solution 1 – Properties of commutators*

(1) *Proof*

a) $[A, B] = AB - BA = -(BA - AB) = -[B, A]$.

b) $[A, (B + C)] = A (B + C) - (B + C) A = AB + AC - BA - CA = (AB - BA) + (AC - CA) = [A, B] + [A, C]$.

c) $[A, B]C + B [A, C] = (AB - BA) C + B(AC - CA) = ABC - BAC + BAC - BCA = (ABC- BCA) = [A, BC]$.

d) $[A, B]^\dagger = (AB - BA)^\dagger = (AB)^\dagger - (BA)^\dagger = B^\dagger A^\dagger - A^\dagger B^\dagger = [B^\dagger, A^\dagger]$.

(2) *Solutions*

a) $(A^\dagger)^\dagger = A$; b) $(\lambda A)^\dagger = \lambda^* A^\dagger$; c) $(\lambda A^\dagger + \mu^* B)^\dagger = (\lambda^* A + \mu B^\dagger$; d) $(AB^\dagger)^\dagger = A^\dagger B$.

(3) *Proof*

a) Let us consider the scalar product:

$$(|\psi\rangle, \left|\frac{d}{dx}\Phi\right\rangle) = \left\langle \psi \left| \frac{d}{dx}\Phi \right.\right\rangle \quad [2.79]$$

Using $\{|x\rangle\}$ representation, the closing relation is written as follows:

$$\int dx |x\rangle\langle x| = \mathbb{1} \quad [2.80]$$

Inserting [2.80] in [2.79], we get:

$$\left\langle \psi \left| \frac{d}{dx}\Phi \right.\right\rangle = \langle\psi| \int_{-\infty}^{+\infty} dx |x\rangle\langle x| \frac{d}{dx}\Phi \rangle$$

It is given as:

$$\left\langle \psi \left| \frac{d}{dx}\Phi \right.\right\rangle = \int_{-\infty}^{+\infty} dx \psi^*(x) \frac{d}{dx}\Phi(x)$$

Using the integration by parts, we have:

$$\left\langle \psi \left| \frac{d}{dx}\Phi \right.\right\rangle = [\psi^* \Phi]_{-\infty}^{+\infty} - \int_{-\infty}^{+\infty} dx \Phi(x) \frac{d}{dx}\psi^*(x)$$

It is given as:

$$\left\langle \psi \left| \frac{d}{dx}\Phi \right.\right\rangle = [\psi^* \Phi]_{-\infty}^{+\infty} - \int_{-\infty}^{+\infty} dx \frac{d}{dx}\psi^*(x)\Phi(x) \quad [2.81]$$

The functions $\psi(x)$ and $\Phi(x)$ related to bound states tend to 0 when $x \rightarrow \pm\infty$, hence the product $\psi^*(x)\,\Phi(x) \rightarrow 0$ when $x \rightarrow \pm\infty$. Therefore, the first term of the right member in [2.81] is zero. Hence:

$$\left\langle \psi \left| \frac{d}{dx}\Phi \right.\right\rangle = -\int_{-\infty}^{+\infty} dx \frac{d}{dx}\psi^*(x)\Phi(x)$$

It is given as:

$$\left\langle \psi \left| \frac{d}{dx}\Phi \right.\right\rangle = -\int_{-\infty}^{+\infty} dx \left(\frac{d}{dx}\psi(x)\right)^* \Phi(x)$$

Or:

$$\left\langle \psi \left| \frac{d}{dx}\Phi \right.\right\rangle = \left\langle -\frac{d}{dx}\psi \left| \Phi \right.\right\rangle \qquad [2.82]$$

Moreover, it is known that if B is the adjoint of A, then the following equality is verified:

$$\langle\psi|A\Phi\rangle = \langle B\psi|\Phi\rangle = \langle A^\dagger\psi|\Phi\rangle \qquad [2.83]$$

with $B = A^\dagger$. We put:

$$A = \frac{d}{dx};\ B = -\frac{d}{dx} \qquad [2.84]$$

Inserting [2.84] into [2.82] and taking [2.83] into account, we get:

$$\left(\frac{d}{dx}\right)^\dagger = -\frac{d}{dx} \qquad [2.85]$$

b) The adjoint of the second derivative operator with respect to variable x is written as:

$$\left(\frac{d^2}{dx^2}\right)^\dagger = \left(\frac{d}{dx}\times\frac{d}{dx}\right)^\dagger = \left(\frac{d}{dx}\right)^\dagger\left(\frac{d}{dx}\right)^\dagger.$$

Using [2.85], we find:

$$\left(\frac{d^2}{dx^2}\right)^{\dagger}=\left(-\frac{d}{dx}\right)^2=\frac{d^2}{dx^2} \qquad [2.86a]$$

c) Using $\{|\vec{r}\rangle\}$ representation, the linear momentum operator is defined by the expression:

$$\vec{P}=-i\hbar\vec{\nabla}=-i\hbar\frac{d}{dx}\vec{i}-i\hbar\frac{d}{dy}\vec{j}-i\hbar\frac{d}{dz}\vec{k} \qquad [2.86b]$$

The adjoint of this operator is written as:

$$\vec{P}^{\dagger}=\left(-i\hbar\frac{d}{dx}\vec{i}-i\hbar\frac{d}{dy}\vec{j}-i\hbar\frac{d}{dz}\vec{k}\right)^{\dagger}$$

It is given as:

$$\vec{P}^{\dagger}=-\left(i\hbar\frac{d}{dx}\vec{i}\right)^{\dagger}-\left(i\hbar\frac{d}{dy}\vec{j}\right)^{\dagger}-\left(i\hbar\frac{d}{dz}\vec{k}\right)^{\dagger}$$

which then gives:

$$\vec{P}^{\dagger}=i\hbar\left(\frac{d}{dx}\vec{i}\right)^{\dagger}+i\hbar\left(\frac{d}{dy}\vec{j}\right)^{\dagger}+i\hbar\left(\frac{d}{dz}\vec{k}\right)^{\dagger} \qquad [2.87a]$$

Taking property [2.85] into account, expression [2.87a] becomes:

$$\vec{P}^{\dagger}=-i\hbar\frac{d}{dx}\vec{i}-i\hbar\frac{d}{dy}\vec{j}-i\hbar\frac{d}{dz}\vec{k} \qquad [2.87b]$$

Comparing [2.86b] and [2.87b], it can be noted that $\vec{P}=\vec{P}^{\dagger}$. This equality proves that the linear momentum operator is Hermitian. It is worth noting that *d*/*dx* is not Hermitian because of the change of sign introduced by the integration by parts (see [2.81] and [2.85]). On the other hand, operator *id*/*dx* is Hermitian due to factor *i*.

2.6.2. *Solution 2 – Trace of an operator*

(1) *Proof*

Let us consider the equation:

$$A\left|u_n^i\right\rangle = a_n\left|u_n^i\right\rangle$$

$$i = 1, 2, 3, \ldots g_n$$

Equation [2.87] is written as:

$$TrA = \sum_n \sum_i^{g_n} \left\langle u_n^i \middle| A \middle| u_n^i \right\rangle$$

This leads to:

$$TrA = \sum_n \sum_i^{g_n} a_n \left\langle u_n^i \middle| u_n^i \right\rangle$$

It is given as:

$$TrA = \sum_n a_n \sum_i^{g_n} \delta_{nn}\delta_{ii}$$

which finally leads to:

$$TrA = \sum_n g_n\, a_n \qquad [2.88]$$

(2) *Proof*

a) By definition:

$$TrAB = \sum_i \langle u_i | AB | u_i \rangle$$

Inserting the closing relation defined in the basis $\{|u_k\rangle\}$ between A and B, we get:

$$TrAB = \sum_i \langle u_i | A \sum_k | u_k \rangle\langle u_k | B | u_i \rangle$$

This leads to:

$$TrAB = \sum_k \sum_i \langle u_k|B|u_i\rangle\langle u_i|A|u_k\rangle$$

which is:

$$TrAB = \sum_k \langle u_k|B\sum_i |u_i\rangle\langle u_i|A|u_k\rangle$$

Hence:

$$TrAB = \sum_k \langle u_k|BA|u_k\rangle = TrBA \qquad [2.89]$$

b) By definition:

$$TrABC = \sum_i \langle u_i|ABC|u_i\rangle$$

As previously, let us insert between operators A and B the closing relation defined in the basis $\{|u_k\rangle\}$ and between B and C the closing relation defined in the basis $\{|u_l\rangle\}$. We get:

$$TrABC = \sum_i \langle u_i|A\sum_k |u_k\rangle\langle u_k|B\sum_l |u_l\rangle\langle u_l|C|u_i\rangle$$

Arranging this expression, we have:

$$TrABC = \sum_k \sum_l \sum_i \langle u_i|A|u_k\rangle\langle u_k|B|u_l\rangle\langle u_l|C|u_i\rangle \qquad [2.90]$$

This leads to:

$$TrABC = \sum_k \sum_l \sum_i \langle u_k|B|u_l\rangle\langle u_l|C|u_i\rangle\langle u_i|A|u_k\rangle$$

Hence:

$$TrABC = \sum_k \langle u_k|B\sum_l |u_l\rangle\langle u_l|C\sum_i |u_i\rangle\langle u_i|A|u_k\rangle$$

which finally leads to:

$$TrABC = \sum_k \langle u_k|BCA|u_k\rangle = TrBCA \qquad [2.91]$$

Using [2.91], we have:

$$TrABC = \sum_k \sum_l \sum_i \langle u_l|C|u_i\rangle\langle u_i|A|u_k\rangle\langle u_k|B|u_l\rangle$$

Or after arrangement:

$$TrABC = \sum_l \langle u_l|CAB|u_l\rangle = TrCAB \qquad [2.92]$$

2.6.3. *Solution 3 – Function of operators*

(1) *Expression*

According to the problem statement, we have:

$$A|\Phi\rangle = a|\Phi\rangle \qquad [2.93]$$

Using [2.93], we get by recurrence:

$$A^2|\Phi\rangle = aA|\Phi\rangle = a^2|\Phi\rangle;\ A^3|\Phi\rangle = a^3|\Phi\rangle;\ldots.\ ;\ A^{n-1}|\Phi\rangle = a^{n-1}|\Phi\rangle$$

or:

$$A^n|\Phi\rangle = a^{n-1}A|\Phi\rangle \Rightarrow A^n|\Phi\rangle = a^n|\Phi\rangle \qquad [2.94]$$

Moreover:

$$F(A)|\Phi\rangle = \sum_n f_n A^n|\Phi\rangle$$

or:

$$F(A)|\Phi\rangle = \sum_n f_n a^n|\Phi\rangle = F(a)|\Phi\rangle \qquad [2.95]$$

(2) *Finding e^{σ_z}*

Using the property [2.95], we find:

$$e^{\sigma_z} = \begin{pmatrix} e & 0 \\ 0 & 1/e \end{pmatrix} \quad [2.96]$$

(3) *Finding the commutators*

Commutators $[A, F(A)]$, $[A, F(B)]$ and $[F(A), F(B)]$ can, respectively, be written as:

$$[A, F(A)] = AF(A) - F(A)A$$

We get:

$$[A, F(A)] = A\sum_n f_n A^n - \sum_n f_n A^n A$$

which is:

$$[A, F(A)] = \sum_n f_n A^{n+1} - \sum_n f_n A^{n+1} = 0 \quad [2.97]$$

According to this result, operator A commutes with any function of A.

$$[A, F(B)] = A\, F(B) - F(B)\, A$$

Similarly:

$$[A, F(B)] = \sum_n g_n AB^n - \sum_n g_n B^n A$$

Factoring by g_n, we get:

$$[A, F(B)] = \sum_n g_n (AB^n - B^n A)$$

which is:

$$[A, F(B)] = \sum_n g_n (AB^n - B^n A) \quad [2.98]$$

If A and B commute, then $AB^n = B^nA \Rightarrow [A, F(B)] = 0$ according to [2.98].

$$[F(A), F(B)] = F(A)F(B) - F(B)F(A)$$

or:

$$[F(A),F(B)] = \sum_n f_n A^n \sum_n g_n B^n - \sum_n g_n B^n \sum_n f_n A^n$$

Since A and B commute, then:

$$AB = BA \Rightarrow A^n B^n = B^n A^n$$

which leads us to:

$$[F(A),F(B)] = \sum_n f_n g_n (A^n B^n - B^n A^n) = 0 \qquad [2.99]$$

2.6.4. Solution 4 – *Infinitesimal unitary operator*

Let us consider an infinitesimal unitary operator $U(\varepsilon)$ defined by the relation:

$$U(\varepsilon) = (\mathbb{1} + A)^{\varepsilon} \qquad [2.100]$$

(1) *Proof*

By definition, an operator is unitary if its adjoint coincides with its inverse. If U is unitary, then:

$$U^{\dagger} = U^{-1} \qquad [2.101]$$

which leads to the relations:

$$U^{\dagger}U = UU^{\dagger} = \mathbb{1} \qquad [2.102]$$

Let U and V be two unitary operators. According to [2.101], we have:

$$U^{\dagger}U = UU^{\dagger} = \mathbb{1};\ V^{\dagger}V = VV^{\dagger} = \mathbb{1} \qquad [2.103]$$

Let us now express the product of operators $(UV)^{\dagger}(UV)$. We get:

$$(UV)^{\dagger}(UV) = (V^{\dagger}U^{\dagger})(UV)$$

Hence:

$$(UV)^{\dagger}(UV) = V^{\dagger}U^{\dagger}UV = V^{\dagger}V = \mathbb{1} \quad [2.104]$$

Result [2.104] shows that the product of two unitary operators is also unitary.

(2) *Expansion to first-order approximation*

In first-order approximation, equation [2.100] gives:

$$U(\varepsilon) = \mathbb{1} + \varepsilon A + \ldots \quad [2.105]$$

(3) *Expression*

Let us express the adjoint of U using [2.105]. We get:

$$U^{\dagger}(\varepsilon) = \mathbb{1} + \varepsilon A^{\dagger} + \ldots \quad [2.106]$$

Using [2.105] and [2.106], the products $U^{\dagger}(\varepsilon)U(\varepsilon)$ and $U(\varepsilon)U^{\dagger}(\varepsilon)$ can be written as:

$$U^{\dagger}(\varepsilon)U(\varepsilon) = \mathbb{1} + \varepsilon(A^{\dagger} + A) + \varepsilon^2 A^{\dagger}A$$

$$U(\varepsilon)U^{\dagger}(\varepsilon) = \mathbb{1} + \varepsilon(A^{\dagger} + A) + \varepsilon^2 AA^{\dagger}$$

Approximating to the first order, we get:

$$U(\varepsilon)U^{\dagger}(\varepsilon) = U^{\dagger}(\varepsilon)U(\varepsilon) = \mathbb{1} + \varepsilon(A^{\dagger} + A) \quad [2.107]$$

(4) *Hermiticity*

Using [2.107], the equation gives:

$$\varepsilon(A^{\dagger} + A) = 0 \Rightarrow A^{\dagger} = -A \quad [2.108]$$

The last equality [2.108] reflects the fact that A is an anti-Hermitian operator.

(5) *Proof*

We put $F(A) = iA$. Taking result [2.108] into account, we get:

$$[F(A)]^{\dagger} = (iA)^{\dagger} = -iA^{\dagger} = iA = F(A) \quad [2.109]$$

Therefore, operator $B = F(A) = iA$ is Hermitian.

2.6.5. Solution 5 – *Properties of Pauli matrices*

We consider the Pauli matrices.

$$\sigma_x = \begin{pmatrix} 0 & 1 \\ 1 & 0 \end{pmatrix} \quad \sigma_y = \begin{pmatrix} 0 & -i \\ i & 0 \end{pmatrix} \quad \sigma_z = \begin{pmatrix} 1 & 0 \\ 0 & -1 \end{pmatrix} \qquad [2.110]$$

The spin angular momentum operator is:

$$\vec{S} = \frac{\hbar}{2}\vec{\sigma} \qquad [2.111]$$

(1) *Proof*

It is known than a Hermitian operator is represented by a Hermitian matrix such that two arbitrary elements that are symmetrical with respect to the main diagonal are complex conjugates. Consequently, the operators represented by matrices [2.111] are self-adjoints or Hermitian.

(2) *Comparison*

We successively calculate the squared operators [2.110] as follows:

$$\sigma_x^2 = \begin{pmatrix} 0 & 1 \\ 1 & 0 \end{pmatrix}\begin{pmatrix} 0 & 1 \\ 1 & 0 \end{pmatrix} = \begin{pmatrix} 1 & 0 \\ 0 & 1 \end{pmatrix}$$

$$\sigma_y^2 = \begin{pmatrix} 0 & -i \\ i & 0 \end{pmatrix}\begin{pmatrix} 0 & -i \\ i & 0 \end{pmatrix} = \begin{pmatrix} 1 & 0 \\ 0 & 1 \end{pmatrix}$$

$$\sigma_z^2 = \begin{pmatrix} 1 & 0 \\ 0 & -1 \end{pmatrix}\begin{pmatrix} 1 & 0 \\ 0 & -1 \end{pmatrix} = \begin{pmatrix} 1 & 0 \\ 0 & 1 \end{pmatrix}$$

These results show that the squared Pauli matrices are all equal to the identity operator I in the space of states with two dimensions. Hence:

$$\sigma_x^2 = \sigma_y^2 = \sigma_z^2 = I = \begin{pmatrix} 1 & 0 \\ 0 & 1 \end{pmatrix} \qquad [2.112]$$

(3) *Expressions of the trace*

Knowing that the *trace of an operator* is the sum of its diagonal matrix elements, using [2.110], it can be seen that:

$$Tr\,(\sigma_x) = Tr\,(\sigma_y) = Tr\,(\sigma_z) = 0 \qquad [2.113]$$

(4) *Proof*

$$\sigma_x\sigma_y + \sigma_y\sigma_x = 0?$$

Using [2110], we get:

$$\sigma_x\sigma_y = \begin{pmatrix} 0 & 1 \\ 1 & 0 \end{pmatrix}\begin{pmatrix} 0 & -i \\ i & 0 \end{pmatrix} = \begin{pmatrix} i & 0 \\ 0 & -i \end{pmatrix}$$

$$\sigma_y\sigma_x = \begin{pmatrix} 0 & -i \\ i & 0 \end{pmatrix}\begin{pmatrix} 0 & 1 \\ 1 & 0 \end{pmatrix} = \begin{pmatrix} -i & 0 \\ 0 & i \end{pmatrix}$$

These expressions prove that:

$$\sigma_x\sigma_y = -\,\sigma_y\sigma_x \Rightarrow \sigma_x\sigma_y + \sigma_y\sigma_x = 0 \qquad [2.114]$$

$$\sigma_x\sigma_y - \sigma_y\sigma_x = 2i\sigma_z?$$

According to [2.114],

$$\sigma_x\sigma_y = -\,\sigma_y\sigma_x \Rightarrow \sigma_x\sigma_y - \sigma_y\sigma_x = -\,2\sigma_y\sigma_x$$

which is:

$$\sigma_x\sigma_y - \sigma_y\sigma_x = 2i^2\sigma_y\sigma_x \qquad [2.115]$$

Let us express the matrix $i^2\sigma_y\sigma_x$. We get:

$$2i^2\sigma_y\sigma_x = 2i^2\begin{pmatrix} 0 & -i \\ i & 0 \end{pmatrix}\begin{pmatrix} 0 & 1 \\ 1 & 0 \end{pmatrix} = 2i^2\begin{pmatrix} -i & 0 \\ 0 & i \end{pmatrix}$$

which is:

$$2i^2\sigma_y\sigma_x = 2i\begin{pmatrix} -i^2 & 0 \\ 0 & i^2 \end{pmatrix} = 2i\begin{pmatrix} 1 & 0 \\ 0 & -1 \end{pmatrix} = 2i\sigma_z \qquad [2.116]$$

After simplification, we find:

$$\sigma_x\sigma_y - \sigma_y\sigma_x = 2i\sigma_z \qquad [2.117]$$

(5) *Expressions of commutators*

Commutators $[\sigma_x, \sigma_y]$, $[\sigma_y, \sigma_z]$ and $[\sigma_z, \sigma_x]$ are written, respectively:

$[\sigma_x, \sigma_y]$

Taking property [2.217] into account, we have:

$$[\sigma_x, \sigma_y] = \sigma_x\sigma_y - \sigma_y\sigma_x \Rightarrow [\sigma_x, \sigma_y] = 2i\sigma_z \qquad [2.118]$$

$[\sigma_y, \sigma_z]$

Proceeding as previously, we get:

$$[\sigma_y, \sigma_z] = \sigma_y\sigma_z - \sigma_z\sigma_y \qquad [2.119]$$

Using [2.110], we get:

$$\sigma_y\sigma_z = \begin{pmatrix} 0 & -i \\ i & 0 \end{pmatrix}\begin{pmatrix} 1 & 0 \\ 0 & -1 \end{pmatrix} = \begin{pmatrix} 0 & i \\ i & 0 \end{pmatrix}$$

$$\sigma_z\sigma_y = \begin{pmatrix} 1 & 0 \\ 0 & -1 \end{pmatrix}\begin{pmatrix} 0 & -i \\ i & 0 \end{pmatrix} = \begin{pmatrix} 0 & -i \\ -i & 0 \end{pmatrix}$$

Hence:

$$\sigma_y\sigma_z = i\begin{pmatrix} 0 & 1 \\ 1 & 0 \end{pmatrix} = i\sigma_x$$

This relation leads to:

$$\sigma_z\sigma_y = -i\begin{pmatrix} 0 & 1 \\ 1 & 0 \end{pmatrix} = -i\sigma_x \qquad [2.120]$$

Using these expressions, relation [2.119] finally gives:

$$[\sigma_y, \sigma_z] = 2i\sigma_x \qquad [2.121]$$

$-[\sigma_z, \sigma_x]$

Similarly:

$$[\sigma_z, \sigma_x] = \sigma_z\sigma_x - \sigma_x\sigma_z \qquad [2.122]$$

Using matrices [2.110], we successively get:

$$\sigma_z\sigma_x = \begin{pmatrix} 1 & 0 \\ 0 & -1 \end{pmatrix}\begin{pmatrix} 0 & 1 \\ 1 & 0 \end{pmatrix} = \begin{pmatrix} 0 & 1 \\ -1 & 0 \end{pmatrix}$$

$$\sigma_x\sigma_z = \begin{pmatrix} 0 & 1 \\ 1 & 0 \end{pmatrix}\begin{pmatrix} 1 & 0 \\ 0 & -1 \end{pmatrix} = \begin{pmatrix} 0 & -1 \\ 1 & 0 \end{pmatrix}$$

which is:

$$\sigma_z\sigma_x = i\begin{pmatrix} 0 & -i \\ i & 0 \end{pmatrix} = i\sigma_y$$

$$\sigma_x\sigma_z = -i\begin{pmatrix} 0 & -i \\ i & 0 \end{pmatrix} = -i\sigma_y$$

Using these results, relation [2.122] finally leads to:

$$[\sigma_z, \sigma_x] = 2i\sigma_y$$

Summarizing the properties of Pauli matrices:

$$\sigma_x^2 = \sigma_y^2 = \sigma_z^2 = I = \begin{pmatrix} 1 & 0 \\ 0 & 1 \end{pmatrix}$$

$$Tr\,(\sigma_x) = Tr\,(\sigma_y) = Tr\,(\sigma_z) = 0 \qquad [2.123]$$

$$\sigma_x\sigma_y + \sigma_y\sigma_x = 0\ ;\ \sigma_x\sigma_y - \sigma_y\sigma_x = 2i\sigma_z$$

$$[\sigma_x, \sigma_y] = 2i\sigma_z\ ;\ [\sigma_y, \sigma_z] = 2i\sigma_x;\ [\sigma_z, \sigma_x] = 2i\sigma_y$$

(6) *Deduction*

Using [2.111] and properties [2.123], we get:

$$S_x = \frac{\hbar}{2}\sigma_x,\ S_y = \frac{\hbar}{2}\sigma_y, S_z = \frac{\hbar}{2}\sigma_z \qquad [2.124]$$

Hence:

$$[S_x, S_y] = \frac{\hbar^2}{2}[\sigma_x, \sigma_y]; [S_y, S_z] = \frac{\hbar^2}{2}[\sigma_y, \sigma_z]; [S_z, S_x] = \frac{\hbar^2}{2}[\sigma_z, \sigma_x]$$

or:

$$[S_x, S_y] = i\hbar S_z \ ; [S_y, S_{yz}] = i\hbar S_x \ ; [S_z, S_x] = i\hbar S_y \qquad [2.125]$$

2.6.6. Solution 6 – *Density operator*

In the discrete orthonormal basis $\{|u_n\rangle\}$, the vector state is written as:

$$|\Psi(t)\rangle = \sum_n c_n(t)|u_n\rangle \qquad [2.126]$$

The density operator is defined by the relation:

$$\rho(t) = |\Psi(t)\rangle\langle\Psi(t)| \qquad [2.127]$$

(1) *Expression*

If ket $|\Psi(t)\rangle$ is normed, then $\langle\Psi(t)|\Psi(t)\rangle = 1$. According to [1.126], we get:

$$\sum_n c_n(t)c_m^*(t)\langle u_m|u_n\rangle = 1$$

Introducing the Kronecker symbol, we have:

$$\sum_n c_n(t)c_m^*(t)\delta_{mn} = 1$$

Or finally:

$$\sum_n |c_n(t)|^2 = 1 \qquad [2.128]$$

(2) *Proof*

Considering [2.127], we get:

$$\rho^\dagger(t) = |\Psi(t)\rangle\langle\Psi(t)| = \rho(t) \qquad [2.129]$$

Relation [2.129] indicates that the density operator is Hermitian.

Furthermore, we calculate the square of this operator. We get:

$$\rho^2(t) = |\Psi(t)\rangle\langle\Psi(t)|\Psi(t)\rangle\langle\Psi(t)| = \rho(t) \qquad [2.130]$$

Operator $\rho(t)$ is therefore idempotent.

(3) *Trace of density operator*

The sum of the diagonal elements of the density matrix is determined by its trace. We then obtain:

$$Tr\rho(t) = \sum_n \langle u_n|\rho(t)|u_n\rangle \qquad [2.131]$$

Inserting [2.127] in [2.131], we get:

$$Tr\rho(t) = \sum_n \langle u_n|\Psi(t)\rangle\langle\Psi(t)|u_n\rangle = \sum_n |\langle u_n|\Psi(t)\rangle|^2$$

Knowing that coefficient $c_n(t) = \langle u_n|\Psi(t)\rangle$, and taking [2.128] into account, we get:

$$Tr\rho(t) = \sum_n |c_n(t)|^2 = 1 \qquad [2.132]$$

2.6.7. *Solution 7 – Evolution operator*

(1) *Relation*

The linear operator $U(k, m)$ is defined by the relation:

$$U(k, m) = |\Phi_k\rangle\langle\Phi_m| \qquad [2.133]$$

Using [2.133], we get:

$$U^\dagger(k, m) = |\Phi_m\rangle\langle\Phi_k| \Rightarrow U^\dagger(k, m) = U(m, k) \qquad [2.134]$$

Let us note that operator $U(k, m)$ is not Hermitian since indices k and m are different.

(2) *Commutator*

The observable A verifies the property:

$$A|\Phi_k\rangle = a_k|\Phi_k\rangle \quad [2.135]$$

Commutator $[A, U(k, m)]$ is:

$$[A, U(k, m)] = AU(k, m) - U(k, m)A$$

Using [2.135], this commutator can be written as:

$$[A, U(k, m)] = A|\Phi_k\rangle\langle\Phi_m| - |\Phi_k\rangle\langle\Phi_m| A$$

Hence:

$$[A, U(k, m)] = a_k|\Phi_k\rangle\langle\Phi_m| - |\Phi_k\rangle\langle\Phi_m| a^*_m$$

Since A is Hermitian (it is an observable), then $a^*_m = a_m$. Hence:

$$[A, U(k, m)] = (a_k - a_m)|\Phi_k\rangle\langle\Phi_m|$$

which is finally given as:

$$[A, U(k, m)] = (a_k - a_m)U(k, m)] \quad [2.136]$$

This expression leads to:

$$[A, U(k, k)] = 0$$

(3) *Proof*

Let us express $Tr\ U(k, m)$ using the basis $\{|\Phi_l\rangle\}$:

$$TrU(k,m) = \sum_l \langle\Phi_l|U(k,m)|\Phi_l\rangle$$

That means using [2.133]:

$$TrU(k,m) = \sum_l \langle\Phi_l|(|\Phi_k\rangle\langle\Phi_m|)|\Phi_l\rangle$$

Hence:

$$TrU(k,m) = \sum_l \langle \Phi_l | \Phi_k \rangle \langle \Phi_m | \Phi_l \rangle = \langle \Phi_m | \left[\sum_l |\Phi_l\rangle\langle\Phi_l| \right] |\Phi_k\rangle$$

So we finally get:

$$Tr\, U\,(k, m) = \langle \Phi_m | \Phi_k \rangle = \delta_{mk} \qquad [2.137]$$

The closing relation satisfied by the set $\{|\Phi_l\rangle\}$ has been used.

(4) *Proof*

The observable B of matrix elements B_{km} can be written as follows:

$$B = \mathbb{1} B \mathbb{1} \qquad [2.138]$$

Using the closing relations verified by the sets $\{|\Phi_k\rangle\}$ and $\{|\Phi_m\rangle\}$, relation [2.138] gives:

$$B = \sum_{k,m} |\Phi_k\rangle\langle\Phi_k| B |\Phi_m\rangle\langle\Phi_m|$$

Hence:

$$B = \sum_{k,m} |\Phi_k\rangle B_{km} \langle\Phi_m| = \sum_{k,m} B_{km} |\Phi_k\rangle\langle\Phi_m|$$

which leads to:

$$B = \sum_{k,m} B_{km} U(k,m) \qquad [2.139]$$

(5) *Proof*

Let us find the trace of $BU^\dagger$. We get:

$$Tr\left[BU^\dagger(k,m)\right] = \sum_l \langle\Phi_l| BU^\dagger(k,m) |\Phi_l\rangle$$

Knowing that $U^\dagger\,(k, m) = |\Phi_m\rangle\langle\Phi_k|$, we get:

$$Tr\left[BU^{\dagger}(k,m)\right] = \sum_{l}\langle\Phi_l|B|\Phi_m\rangle\langle\Phi_k|\Phi_l\rangle$$

Hence:

$$Tr\left[BU^{\dagger}(k,m)\right] = \sum_{l}B_{lm}\langle\Phi_k|\Phi_l\rangle$$

So we finally get:

$$B_{km} = Tr\{BU^{\dagger}(k,m)\} \quad [2.140]$$

(6) *Evolution equation*

$$|\Psi(t)\rangle = U(t,t_0)|\Psi(t_0)\rangle \quad [2.141]$$

In this equation, $U(t, t_0)$ is the evolution operator.

(6.1) *Proof*

Let us express the norm of ket vector $|\Psi(t)\rangle$ using [2.141]. We get:

$$\langle\Psi(t)|\Psi(t)\rangle = \langle\Psi\, t_0)|\, U^{\dagger}(t,t_0)\, U(t,t_0)|\Psi(t_0)\rangle$$

Knowing that $\langle\Psi(t)|\Psi(t)\rangle = 1$, we have:

$$U^{\dagger}(t,t_0)\, U(t,t_0)\langle\Psi\, t_0)|\Psi(t_0)\rangle = 1$$

which finally leads to:

$$U^{\dagger}(t,t_0)\, U(t,t_0) = \mathbb{1} \quad [2.142]$$

Result [2.142] indicates that $U^{\dagger}(t,t_0) = U^{-1}(t,t_0)$: $U(t, t_0)$ is therefore unitary.

(6.2) *Proof*

The Schrödinger equation that governs the evolution of the conservative system is:

$$i\hbar\frac{d|\Psi(t,t_0)\rangle}{dt} = H|\Psi(t,t_0)\rangle \quad [2.143]$$

Inserting [2.141] in [2.143], we get:

$$i\hbar\frac{dU(t,t_0)|\Psi(t_0)\rangle}{dt} = HU(t,t_0)|\Psi(t_0)\rangle$$

Hence:

$$i\hbar\frac{dU(t,t_0)}{dt} = HU(t,t_0) \qquad [2.144]$$

(6.3) *Expression*

As the Hamiltonian H is time independent (the system is conservative), expression [6.220] leads to the relation, which involves the variables t and t_0:

$$\frac{dU(t,t_0)}{U(t,t_0)} = \frac{H}{i\hbar}dt$$

Or after integration:

$$U(t,t_0) = e^{-iH(t,t_0)/\hbar} \qquad [2.145]$$

Since $U(t_0, t_0) = \mathbb{1}$

2.6.8. *Solution 8 – Orbital angular momentum operator*

Orbital angular momentum operator with respect to the origin point O is:

$$\vec{l} = \vec{r} \wedge \vec{p} \qquad [2.146]$$

The component p_q of the linear momentum operator is given by the relation:

$$p_q = -i\hbar\frac{\partial}{\partial q} \qquad [2.147]$$

(1) *Expressions of components*

Using [2.146], operators l_x, l_y and l_z are written, respectively:

$$\begin{cases} l_x = (yp_z - zp_y) = i\hbar\left(z\dfrac{\partial}{\partial y} - y\dfrac{\partial}{\partial z}\right) \\ l_y = (zp_x - xp_z) = i\hbar\left(x\dfrac{\partial}{\partial z} - z\dfrac{\partial}{\partial x}\right) \\ l_z = (xp_y - yp_x) = i\hbar\left(y\dfrac{\partial}{\partial x} - x\dfrac{\partial}{\partial y}\right) \end{cases} \quad [2.148]$$

(2) *Expressions of products*

Considering [2.148], we have:

$$L_xL_y = -\hbar^2\left(z\frac{\partial}{\partial y} - y\frac{\partial}{\partial z}\right)\left(x\frac{\partial}{\partial z} - z\frac{\partial}{\partial x}\right)$$

That means:

$$L_xL_y = -\hbar^2\left(z\frac{\partial}{\partial y}x\frac{\partial}{\partial z} - y\frac{\partial}{\partial z}x\frac{\partial}{\partial z}\right)\left(-z\frac{\partial}{\partial y}z\frac{\partial}{\partial x} + y\frac{\partial}{\partial z}z\frac{\partial}{\partial x}\right)$$

After expansion, we find:

$$L_xL_y = -\hbar^2\left(zx\frac{\partial^2}{\partial y\partial z} - yx\frac{\partial^2}{\partial z^2} - z^2\frac{\partial^2}{\partial y\partial x} + yz\frac{\partial^2}{\partial z\partial x} + y\frac{\partial}{\partial x}\right) \quad [2.149]$$

Using the same reasoning as previously, we find:

$$L_yL_x = -\hbar^2\left(zy\frac{\partial^2}{\partial x\partial z} - xy\frac{\partial^2}{\partial z^2} - z^2\frac{\partial^2}{\partial x\partial y} + xz\frac{\partial^2}{\partial z\partial y} + x\frac{\partial}{\partial y}\right) \quad [2.150]$$

It is worth noting that partial derivatives with respect to two independent variables are commutative, hence:

$$\frac{\partial^2}{\partial x\partial y} = \frac{\partial^2}{\partial y\partial x} \quad [2.151]$$

(3) *Commutators*

Subtracting the two equalities [2.149] and [2.150], member by member, we find after simplification and arrangement:

$$l_x l_y - l_y l_x = i\hbar\, l_z$$

Circular permutation can be used to obtain the other commutation relations:

$$l_z l_x - l_x l_z = i\hbar\, l_y$$

$$l_y l_z - l_z l_y = i\hbar\, l_x$$

Summarizing, we have:

$$[l_x, l_y] = i\hbar\, l_z \;;\; [l_y, l_z] = i\hbar\, l_x \;;\; [l_z, l_x] = i\hbar\, l_y \qquad [2.152]$$

NOTE.– Relations [2.152] express that two arbitrary components of the angular momentum operator are not commutative. Consequently, there is no physical state in which the three or two projections of the angular momentum have determined values. In other words, there is no physical state in which the magnitude and direction of the angular momentum are fully determined, contrary to the predictions of classical mechanics. Only the squared angular momentum operator and one of its projections can be simultaneously determined (see Exercise 3.5.4, Chapter 3).

3

Eigenvalues and Eigenvectors of an Observable

General objective

The general objective is to know the properties of the eigenvalues and eigenvectors of an observable.

Specific objectives

On completing this chapter, the reader should be able to:

– define a representation;

– represent a ket and a bra;

– represent an operator;

– represent the adjoint of an operator;

– recognize a Hermitian matrix;

– determine the properties of the eigenvalues of an observable;

– determine the properties of the eigenvectors of an observable;

– distinguish between a simple eigenvalue and a degenerate eigenvalue;

– use the characteristic equation;

– know the properties of the eigenvectors and eigenvalues of a Hermitian operator;

– establish the evolution equation of the mean value of an observable;

For color versions of the figures in this book, see www.iste.co.uk/sakho/quantum2.zip

– define a complete set of commuting observables (CSCO);

– know the properties of conservative systems;

– integrate Schrödinger's equation applied to conservative systems;

– establish Ehrenfest's theorem.

Specific objectives

– Matrix calculus.

– Observable.

– Hamiltonian.

– Properties of the space of states.

3.1. Representation

3.1.1. *Definition*

In quantum mechanics, the passage from *vector calculus* introduced in the space of states to *matrix calculus* in the same space is based on the choice of a *representation*. For this purpose, a discrete or continuous orthonormal basis is chosen, in which:

– kets and bras are represented by numbers (their components on the basis vectors);

– operators are represented by their matrix elements.

This study focuses on the case of discrete bases $\{|u_i\rangle\}$, $i = 1, 2, 3, \ldots$. It requires using the orthonormalization relations [2.17] and the closing relations [2.21], as well as the relation of definition of a matrix element [2.25] between kets $|\Phi\rangle$ and $|\Psi\rangle$ which are summarized as follows:

$$\langle u_i|u_j\rangle = \delta_{ij}\,;\; c_i = \langle u_i|\Psi\rangle\,;\; \sum_i |u_i\rangle\langle u_i| = \mathbb{1}\,;\; \langle\Phi|A|\Psi\rangle \qquad [3.1]$$

As will be seen in section 3.2, the choice of a representation in the space of states makes it possible to determine the eigenvectors and eigenvalues of a given observable based on operations on matrices.

3.1.2. *Representation of kets and bras*

By convention,

– a ket is represented by a *single column matrix* with a countable infinity of rows. The *matrix elements* being the components c_i given by the second of relations [2.20];

– a bra is represented by a *single row matrix* with a countable infinity of columns. The matrix elements are complex conjugates c^*_i of the components c_i of the state vector on the chosen basis, hence: $c_i^* = \langle\Psi|u_i\rangle$.

The representations of ket $|\Psi\rangle$ and bra $\langle\Psi|$ are given below (Figure 3.1).

$$\begin{pmatrix} \langle u_1|\Psi\rangle \\ \langle u_2|\Psi\rangle \\ \cdots\cdots \\ \cdots\cdots \\ \langle u_i|\Psi\rangle \\ \cdots\cdots \\ \cdots\cdots \\ \cdots\cdots \end{pmatrix} \text{ or } \begin{pmatrix} c_1 \\ c_2 \\ \cdots\cdots \\ \cdots\cdots \\ c_i \\ \cdots\cdots \\ \cdots\cdots \\ \cdots\cdots \end{pmatrix}$$

a)

$$\begin{pmatrix} \langle\Psi|u_1\rangle & \langle\Psi|u_2\rangle & \ldots & \langle\Psi|u_i\rangle & \ldots \end{pmatrix}$$

or

$$\begin{pmatrix} c_1 & c_2 & \ldots\ldots & c_i^* & \ldots\ldots \end{pmatrix}$$

b)

Figure 3.1. *Representation of a) kets and b) bras in the space of states*

3.1.3. *Representation of operators*

This section covers only the linear operators.

Let us consider the discrete set $\{|u_i\rangle\}$. The *matrix elements* denoted A_{ij} of an operator A are defined by the expression:

$$A_{ij} = \langle u_i|A|u_j\rangle. \qquad [3.2]$$

Operator A is represented by a *square matrix* $N \times N$ of elements A_{ij}. It is a matrix with N rows and N columns; index i identifies the row, while index j identifies the column. The representation shown in Figure 3.2 is thus obtained.

$$(A) = \begin{pmatrix} A_{11} & A_{12} & A_{13} & . & . & A_{1j} & . \\ A_{21} & A_{22} & A_{23} & . & . & A_{2j} & . \\ A_{31} & A_{32} & A_{33} & . & . & A_{3j} & . \\ . & . & . & . & . & . & . \\ . & . & . & . & . & . & \\ A_{i1} & A_{i2} & A_{i3} & . & . & A_{ij} & . \\ . & . & . & . & . & . & . \\ . & . & . & . & . & . & . \end{pmatrix}$$

Figure 3.2. *Representation of an operator A*

APPLICATION **3.1.** –

We represent the projection operator $P_\psi = |\psi\rangle\langle\psi|$ in the basis $\{|u_i\rangle\}$, $i = 1, 2$.

Solution. In the two-dimensional basis $\{|u_i\rangle\}$, vectors ket $|\psi\rangle$ and bra $\langle\psi|$ are provided by the expansions:

$$\begin{cases} |\psi\rangle = \sum_{i=1}^{2} c_i |u_i\rangle = c_1|u_1\rangle + c_2|u_2\rangle \\ \langle\psi| = \sum_{i=1}^{2} c_i^* \langle u_i| = c_1^*\langle u_1| + c_2^*\langle u_2| \end{cases} \quad [3.3]$$

Using expansions [3.3], the representations of ket $|\psi\rangle$ and bra $\langle\psi|$ can be written according to Figure 3.1:

$$(|\psi\rangle) = \begin{pmatrix} c_1 \\ c_2 \end{pmatrix}; \quad (\langle\psi|) = \begin{pmatrix} c_1^* & c_2^* \end{pmatrix} \quad [3.4]$$

The projection vector $P_\psi = |\psi\rangle\langle\psi|$ is then represented by the following square matrix 2×2:

$$(P_\psi) = \begin{pmatrix} c_1 \\ c_2 \end{pmatrix} \begin{pmatrix} c_1^* & c_2^* \end{pmatrix} \Rightarrow (P_\psi) = \begin{pmatrix} c_1c_1^* & c_1c_2^* \\ c_2c_1^* & c_2c_2^* \end{pmatrix}. \qquad [3.5]$$

3.1.4. *Hermitian matrix*

As already mentioned, an operator A is Hermitian if it coincides with its adjoint $A^\dagger$. It is important to be able to recognize if A is Hermitian based on its matrix representation. For this purpose, we first need to state the principle for the *representation of the adjoint of an operator.*

Let us thus consider the complex conjugate of the matrix elements A_{ij} of operator A defined by relation [3.2]. We have:

$$(A_{ij})^* = (\langle u_i|A|u_j\rangle)^* = \langle u_j|A^\dagger|u_i\rangle = (A^\dagger)_{ji} \qquad [3.6]$$

According to [3.6], the adjoint $A^\dagger$ of operator A is represented by a *square matrix* $N \times N$ of *matrix elements* $(A^\dagger)_{ji} = (A_{ij})^*$. In order to represent $A^\dagger$, we consider the complex conjugates of matrix elements A_{ij} of A followed by an inversion with respect to the *main diagonal.*

As an illustration, let us represent the adjoint $A^\dagger$ of operator A represented by the first matrix below. We have:

$$(A) = \begin{pmatrix} i & 2i \\ 3i & 1 \end{pmatrix} \;;\; (A)^* = \begin{pmatrix} -i & -2i \\ -3i & 1 \end{pmatrix} \;;\; (A^\dagger) = \begin{pmatrix} -i & -3i \\ -2i & 1 \end{pmatrix} \qquad [3.7]$$

Let us now consider the specific case of Hermitian operators.

If A is Hermitian, then $A = A^\dagger$. Hence according to [3.6]:

$$(A_{ij})^* = (A)_{ji} \text{ and } (A_{ii})^* = (A)_{ii} \qquad [3.8]$$

CONCLUSION.– A Hermitian operator is represented by a *Hermitian matrix* such that the *diagonal matrix elements* are *real.* Moreover, two arbitrary *matrix elements symmetrical* with respect to the *main diagonal* are complex conjugates.

APPLICATION 3.2.–

Represent the adjoints $A^\dagger$, $B^\dagger$ and $C^\dagger$ of operators A, B and C represented below. Identify those that are Hermitian.

$$(A)=\begin{pmatrix}0 & 2\\ i & 0\end{pmatrix} \;;\; (B)=\begin{pmatrix}-2i & -i\\ i & i\end{pmatrix} \;;\; (C)=\begin{pmatrix}1 & -i\\ i & 1\end{pmatrix} \qquad [3.9]$$

Solution. We have:

$$(A)^\dagger=\begin{pmatrix}0 & -i\\ 2 & 0\end{pmatrix} \;;\; (B)^\dagger=\begin{pmatrix}-2i & -i\\ i & -i\end{pmatrix} \;;\; (C)^\dagger=\begin{pmatrix}1 & -i\\ i & 1\end{pmatrix} \qquad [3.10]$$

According to [3.10], only C is Hermitian. B is not Hermitian because its diagonal matrix elements are not real.

3.2. Eigenvalues equation, mean value

3.2.1. *Definitions, degeneracy*

As already noted, a linear operator establishes a correspondence between any ket $|\Psi\rangle$ in the space of states E_r, and another ket $|\Psi'\rangle$ belonging to the same space such that $A|\Psi\rangle = |\Psi'\rangle$.

Let us consider the specific case when the initial ket $A|\psi\rangle$ is proportional to ket $|\psi\rangle$. Let λ be the coefficient of proportionality. We have:

$$A|\psi\rangle = \lambda|\psi\rangle \qquad [3.11]$$

By definition, equation [3.11] is known as *equation with eigenvalues* of operator A. In this equation, λ designates the *eigenvalue* of operator A and $|\Psi\rangle$ represents the *eigenket* or the *eigenvector* of operator A associated with the eigenvalue λ.

An important example is that of the Hamiltonian H of a system whose total energy is E. The equation with eigenvalues can be written as: $H|\psi\rangle = E|\psi\rangle$.

When equation [3.11] is considered, two cases can be distinguished.

First case: The eigenvalue λ is non-degenerate (or simple).

The eigenvalue λ is known as *non-degenerate* or *simple* if, ignoring a multiplicative factor, only one eigenvector or ket is associated with it.

A simple example is the case of a particle of mass m confined in a potential well of infinite depth and width a. The eigenvalue of the Hamiltonian is quantized, and is given by the expression [1.63], which is:

$$E_n = \frac{\hbar^2\pi^2}{2ma^2}n^2;\ E_0 = \frac{\hbar^2\pi^2}{2ma^2}n^2$$

or:

$$E_n = E_0 n^2 \qquad [3.12]$$

The eigenvalue equation can be written as follows:

$$H|\psi_n\rangle = E_n|\psi_n\rangle \qquad [3.13]$$

– for the ket $|\psi_1\rangle$: $E_1 = E_0$;

– for the ket $|\psi_2\rangle$: $E_2 = 4E_0$;

– for the ket $|\psi_3\rangle$: $E_3 = 9E_0$;

– ....;

– for the ket $|\psi_n\rangle$: $E_n = n^2E_0$.

These results actually show that the eigenvalue E_n is non-degenerate since it is associated with a single ket $|\psi_n\rangle$.

Second case: The eigenvalue λ is degenerate.

The *eigenvalue* λ is known as *degenerate* if it is associated with two or more eigenvectors or kets.

A simple example is that of a particle of mass m confined in a square potential well. The eigenvalue of the Hamiltonian is given by the expression [1.223], which is recalled below:

$$E'_{n,q} = E_0\,(n^2 + q^2)$$

The equation with eigenvalues can be written in the following form:

$$H|\psi_{n,q}\rangle = E_{n,q}|\psi_{n,q}\rangle \qquad [3.14]$$

Using [3.14], we have:

– for the ket $|\psi_{1,1}\rangle$: $E_{1,1} = 2E_0$; non-degenerate eigenvalue $E_{1,1}$;

– for the ket $|\psi_{1,2}\rangle$: $E_{1,2} = 5E_0$;

– for the ket $|\psi_{2,1}\rangle$: $E_{2,1} = 5E_0$.

The eigenvalue $E_{2,1}$ or $E_{1,2}$ is therefore *twice degenerate* since two different kets $|\psi_{1,2}\rangle$ and $|\psi_{2,1}\rangle$ are associated with it. The same is true for all the eigenvalues $E_{n,q}$ when $n \neq q$.

As previously explained, this degeneracy is due to the symmetry of the potential. For a rectangular well, the eigenvalues $E_{n,q}$ are all simple, as shown by the expression [1.222].

Therefore, in the general case, the eigenvalue λ is g_n times degenerate if it is associated with g_n different eigenkets. In this case, the eigenvectors of A form a *vector subspace* of the eigenvalue λ of dimension g_n.

Denoting by $\left|u_n^i\right\rangle$ the eigenvalues of A, the vector subspace of the eigenvalue λ is then subtended by the set $\{\left|u_n^i\right\rangle\}$, $i = 1, 2, 3,\ldots, g_n$. The expansion of the state vector is then:

$$|\Psi\rangle = \sum_{n=1}^{\infty}\sum_{i=1}^{g_n} c_n^i \left|u_n^i\right\rangle \tag{3.15}$$

APPLICATION 3.3. –

Let $\lambda = a_n$ be the eigenvalues of an operator A associated with the eigenvectors $\left|u_n^i\right\rangle$. What can be said about the following ket:

$$|\psi_i\rangle = \sum_{i=1}^{g_n} c_n^i \left|u_n^i\right\rangle \tag{3.16}$$

Solution. The equation with eigenvalues of A is:

$$A\left|u_n^i\right\rangle = a_n\left|u_n^i\right\rangle \Rightarrow A|\psi_i\rangle = \sum_{i=1}^{g_n} c_n^i A\left|u_n^i\right\rangle \tag{3.17}$$

Knowing that summing is not applicable over n, we have:

$$A|\psi_i\rangle = a_n \sum_{i=1}^{g_n} c_n^i |u_n^i\rangle \Rightarrow A|\psi_i\rangle = a_n|\psi_i\rangle \qquad [3.18]$$

According to [3.18], $|\psi_i\rangle$ is an eigenket of A associated with the same eigenvalue a_n.

3.2.2. Characteristic equation

Equation [3.11] expresses the relation between the eigenvalue λ and the eigenvector $|\psi\rangle$ of the observable A. The issue is to find out how to determine λ and $|\psi\rangle$. In what follows, we establish an equation that makes it possible to determine the eigenvalues of an observable. Once these eigenvalues are known, it becomes easy to deduce from them the associated eigenvectors using equation [3.11].

Let $\{|u_i\rangle\}$ be a discrete orthonormal set. Projecting the equation with eigenvalues [3.11] on the ket $|u_i\rangle$, we obtain:

$$\langle u_i|A|\psi\rangle = \lambda \langle u_i|\psi\rangle \qquad [3.19]$$

Inserting the closing relation verified by the set $\{|u_j\rangle\}$ in the right member of equation [3.19], we have:

$$\langle u_i|A\sum_j |u_j\rangle\langle u_j||\psi\rangle = \lambda\langle u_i|\psi\rangle$$

This yields:

$$\sum_j \langle u_i|A|u_j\rangle\langle u_j|\psi\rangle = \lambda\langle u_i|\psi\rangle$$

which is:

$$\sum_j A_{ij}c_j = \lambda c_i \qquad [3.20]$$

Knowing that (see proof leading to relation [2.19]):

$$c_i = \sum_j c_j \delta_{ij}$$

we have:

$$\sum_j A_{ij} c_j = \lambda \sum_j c_j \delta_{ij}$$

which is:

$$\sum_j \left(A_{ij} - \lambda \delta_{ij}\right) c_j = 0 \qquad [3.21]$$

Relation [3.21] can be considered as a system of equations whose unknowns are the components c_j of the eigenvector $|\psi\rangle$ on the basis $\{|u_j\rangle\}$. It is a homogeneous and linear system with N unknowns c_j (j = 1, 2, ..., N) involving N equations (i = 1, 2, ... N). Because the system [3.21] is homogeneous and linear, it has one solution (other than the trivial solution for which all c_j are zero) if and only if the *determinant* of the coefficients is zero, which means:

$$\text{Det}\,[A - \lambda I] = 0 \qquad [3.22]$$

Equation [3.22] is known as a *characteristic equation* or *secular equation*. In this equation, A is a square matrix $N \times N$ of matrix elements A_{ij}, and I designates the *unit matrix*.

APPLICATION 3.4. –

Determine the eigenvalues and the eigenvectors of operator A represented by the following matrix:

$$(A) = \begin{pmatrix} 1 & 2i \\ -i & 0 \end{pmatrix} \qquad [3.23]$$

Solution.

1) *Eigenvalues of A*

The matrix of A is a square matrix 2 × 2. Let $|\psi\rangle$ be the eigenvector of A associated with the eigenvalue λ. Using the secular equation [3.22] and the matrix [3.23], we obtain:

$$\text{Det}\,(\mathcal{A} - \lambda I) = \begin{pmatrix} 1-\lambda & 2i \\ -i & -\lambda \end{pmatrix} = 0 \qquad [3.24]$$

Then this equation is written as follows:

$-\lambda(1-\lambda) - 2 = 0 \Rightarrow \lambda_1 = -1$ and $\lambda_2 = 2$

2. *Eigenvectors of A*

The equation with eigenvalues of A is given by [3.11]. Equation [3.24] provides two eigenvalues of operator A. Let $|\psi_1\rangle$ and $|\psi_2\rangle$ be, respectively, the associated eigenvectors. The ket expansion in the orthonormal basis $\{|u_i\rangle\}$ is written as follows:

$$|\psi\rangle = \sum_{i=1}^{2} c_i |u_i\rangle = c_1|u_1\rangle + c_2|u_2\rangle \qquad [3.25]$$

According to the condition for the normalization of $|\psi\rangle$, we have:

$$\langle\psi|\psi\rangle = 1 \Rightarrow \sum_{i=1}^{2} |c_i|^2 = |c_1|^2 + |c_2|^2 = 1 \qquad [3.26]$$

In the orthonormal basis $\{|u_i\rangle\}$, the matrix equation corresponding to the equation with eigenvalues [3.11] is written using [3.23]:

$$\begin{pmatrix} 1 & 2i \\ -i & 0 \end{pmatrix}\begin{pmatrix} c_1 \\ c_2 \end{pmatrix} = \lambda \begin{pmatrix} c_1 \\ c_2 \end{pmatrix}$$

This leads to:

$$\begin{cases} c_1 + 2ic_2 = \lambda c_1 \\ -ic_1 \quad\;\; = \lambda c_2 \end{cases} \qquad [3.27]$$

– *Case of eigenvalue* $\lambda_1 = -1$

The second equation of system [3.27] yields: $ic_1 = c_2$. Using [3.26] and making an arbitrary choice for a real and positive c_1, we get:

$$|c_1|^2 + |ic_1|^2 = 1 \Rightarrow c_1 = \frac{1}{\sqrt{2}}\ ;\ c_2 = \frac{i}{\sqrt{2}} \qquad [3.28]$$

The eigenvector of A is then written using the expansion [3.25]:

$$\left|\psi_1\right\rangle=\frac{1}{\sqrt{2}}\left|u_1\right\rangle+\frac{i}{\sqrt{2}}\left|u_2\right\rangle \quad [3.29]$$

– *Case of the eigenvalue* $\lambda_2 = 2$

Using the same reasoning as previously, the eigenvector of A associated with the eigenvalue $\lambda_2 = 2$ can be written as:

$$\left|\psi_2\right\rangle=\frac{2i}{\sqrt{5}}\left|u_1\right\rangle+\frac{1}{\sqrt{5}}\left|u_2\right\rangle \quad [3.30]$$

3.2.3. *Properties of eigenvectors and eigenvalues of a Hermitian operator*

The *eigenvectors* and *eigenvalues* of a Hermitian operator verify the following two fundamental properties.

First property: The eigenvalues of a Hermitian operator are real.

Proof.

Let us consider the equation with eigenvalues [3.11]. We project this equation onto the ket $|\psi\rangle$. We obtain:

$$\langle\psi|A|\psi\rangle = \lambda\langle\psi|\psi\rangle = \lambda \quad [3.31]$$

The complex conjugate of [3.31] can be written knowing that A is Hermitian:

$$(\langle\psi|A|\psi\rangle)^* = \lambda^* \Rightarrow \langle\psi|A^\dagger|\psi\rangle = \langle\psi|A|\psi\rangle = \lambda^* \quad [3.32]$$

Comparing [3.31] and [3.32], we see that $\lambda^* = \lambda$. The property is therefore proved.

Second property: Two eigenvectors of a Hermitian operator associated with two different eigenvalues are orthogonal.

Proof.

Let λ_1 and λ_2 be two eigenvalues of a Hermitian operator A, associated with the respective eigenvectors $|\psi_1\rangle$ and $|\psi_2\rangle$. The equations with the corresponding eigenvalues are written according to [3.11]:

$$A|\psi_1\rangle = \lambda_1|\psi_1\rangle;\ A|\psi_2\rangle = \lambda_2|\psi_2\rangle \qquad [3.33]$$

We project the first of equations [3.33] on the ket $|\psi_2\rangle$ and the second on the ket $|\psi_1\rangle$. We obtain:

$$\langle\psi_2|A|\psi_1\rangle = \lambda_1\langle\psi_2|\psi_1\rangle;\ \langle\psi_1|A|\psi_2\rangle = \lambda_2\langle\psi_1|\psi_2\rangle \qquad [3.34]$$

Knowing that λ_1 and λ_2 are real and A is Hermitian, the complex conjugate of the first equation [3.34] is then written as follows:

$$\langle\psi_1|A|\psi_2\rangle = \lambda_1\langle\psi_1|\psi_2\rangle \qquad [3.35]$$

Equalizing [3.35] and the second equation [3.34], and knowing that $\lambda_1 \neq \lambda_2$, we have:

$$(\lambda_1 - \lambda_2)\,\langle\psi_1|\psi_2\rangle = 0 \Rightarrow \langle\psi_1|\psi_2\rangle = 0 \qquad [3.36]$$

The result [3.36] reflects the orthogonality of eigenvectors $|\psi_1\rangle$ and $|\psi_2\rangle$.

3.2.4. *Evolution of the mean value of an observable*

Let $|\Psi(t)\rangle$ be a ket normed in the space of states; by definition, the *mean value of an observable* A (t) denoted by $\langle A(t)\rangle$ is given by the relation:

$$\langle A(t)\rangle = \langle\Psi(t)|A(t)|\Psi(t)\rangle \qquad [3.37]$$

If the state vector $|\Psi(t)\rangle$ is not normed, then [3.37] should be divided by the squared norm $\langle\Psi(t)|\Psi(t)\rangle$. Let us differentiate $\langle A(t)\rangle$ with respect to time taking into account the fact that the observable A may depend on other quantities, such as the position (case of potential $V(r, t)$). We obtain:

$$\frac{d\langle A(t)\rangle}{dt} = \left[\frac{d}{dt}\langle\Psi(t)|\right]A(t)|\Psi(t)\rangle + \langle\Psi(t)|\frac{\partial A(t)}{\partial t}|\Psi(t)\rangle + \langle\Psi(t)|A(t)\left[\frac{d}{dt}|\Psi(t)\rangle\right] \qquad [3.38]$$

Using Schrödinger's equation [1.20], we have:

$$i\hbar\frac{d|\Psi(t)\rangle}{dt} = H(t)|\Psi(t)\rangle \Rightarrow -i\hbar\frac{d\langle\Psi(t)|}{dt} = \langle\Psi(t)|H(t) \qquad [3.39]$$

Using [3.39], the evolution equation [3.38] is written as:

$$\frac{d\langle A(t)\rangle}{dt} = -\frac{\langle\Psi(t)|HA|\Psi(t)\rangle}{i\hbar} + \langle\Psi(t)|\frac{\partial A(t)}{\partial t}|\Psi(t)\rangle + \frac{\langle\Psi(t)|AH|\Psi(t)\rangle}{i\hbar}$$

Hence:

$$\frac{d\langle A(t)\rangle}{dt} = \frac{1}{i\hbar}\left[\langle\Psi(t)|AH|\Psi(t)\rangle - \langle\Psi(t)|HA|\Psi(t)\rangle\right] + \langle\Psi(t)|\frac{\partial A(t)}{\partial t}|\Psi(t)\rangle$$
$$= \frac{1}{i\hbar}\langle\Psi(t)|\left[AH - HA\right]|\Psi(t)\rangle + \langle\Psi(t)|\frac{\partial A(t)}{\partial t}|\Psi(t)\rangle$$

Finally:

$$\frac{d\langle A(t)\rangle}{dt} = \frac{1}{i\hbar}\langle[A,H]\rangle + \left\langle\frac{\partial A(t)}{\partial t}\right\rangle \qquad [3.40]$$

Relation [3.40] expresses the *equation of evolution* of the mean value $\langle A(t)\rangle$ of the observable $A(t)$. This equation involves two remarks.

If the two terms of the right member of equation [3.40] are simultaneously equal to zero, then $\langle A(t)\rangle$ = constant. Hence: $d\langle A(t)\rangle/dt = 0$. Let us put:

$$\begin{cases} [A,H]=0 \\ \dfrac{\partial A(t)}{\partial t}=0 \end{cases} \qquad [3.41]$$

An arbitrary observable that simultaneously satisfies the two equations [3.41] is known as a *constant of motion*.

By definition, an observable A is a constant of motion if:

1) It commutes with the Hamiltonian H;

2) It does not explicitly depend on time.

The Hamiltonian H of a conservative system is an example of a constant of motion (see section 3.3).

3.2.5. *Complete set of commuting observables*

Throughout this chapter, it has been shown that Hermitian operators play a very important role in quantum mechanics due to the possibility to build a basis in the space of states from their eigenvectors. Moreover, a constant of motion such as the observable A commutes with the Hamiltonian H. There is a particularly interesting case in which one can choose a single basis (ignoring a multiplicative factor) constituted of eigenvectors common to observables A and H. In this particular case, the set A and H constitutes a *complete set of commuting observables* (CSCO).

In the general case, a set of observables A, B, C,......, is known as complete set of commuting observables if [COH 77]:

– all the observables A, B, C,..., commute;

– given the eigenvectors of A, B, C,..., it is sufficient to define a single basis (ignoring a multiplicative factor) that is common to the set of observables.

The notion of CSCO plays a very important role in quantum mechanics. It is, for example, the case of conservative systems, where a basis can be built in the space of states from eigenvectors common to the Hamiltonian H, to the component L_z and to the square $\vec{l}^{\,2}$ of the angular momentum operator defined in Chapter 2 (Volume 2).

3.3. Conservative systems

3.3.1. *Definition*

A *conservative system* is a system whose Hamiltonian does not depend on time: $H(t)$ = constant. Or, according to [3.40], putting $A(t) = H(t)$:

$$\frac{d\langle H(t)\rangle}{dt} = \frac{1}{i\hbar}\langle [H,H]\rangle + \left\langle \frac{\partial H(t)}{\partial t} \right\rangle = 0 \qquad [3.42]$$

This equation shows that the Hamiltonian of conservative systems is actually a constant of motion. The fact that H is a constant of motion entails the conservation of the total energy E. This corresponds to the classical case of the system not being subjected to any friction force.

Moreover, the states of the conservative system are stationary states and its evolution is governed by the time-independent Schrödinger equation [1.29]. Many conservative systems have been studied in Chapter 1. It is the case of a particle confined in a potential well, a quantum harmonic oscillator, the hydrogen atom, etc.

3.3.2. *Integration of Schrödinger's equation*

Let $\{|\varphi_{n,k}\rangle\}$ be a set of eigenkets of the Hamiltonian H of a conservative system. The equation with eigenvalues is written as follows:

$$H|\varphi_{n,k}\rangle = E_n|\varphi_{n,k}\rangle \qquad [3.43]$$

In the equation [3.43], E_n is the eigenvalue of the Hamiltonian H and the index k identifies the eigenstates of the observables that, together with H, constitute a CSCO.

Let us consider the simple case when the basis $\{|\varphi_{n,k}\rangle\}$ is common to the set H and A. The equation with eigenvalues of the observable A of eigenvalues a_k is then written as follows:

$$A|\varphi_{n,k}\rangle = a_k|\varphi_{n,k}\rangle \qquad [3.44]$$

The state vector is given by the expression:

$$|\Psi(t)\rangle = \sum_{n,k}^{\infty} c_{n,k}(t)|\varphi_{n,k}\rangle \qquad [3.45]$$

In [3.45], the components $c_{n,k}(t)$ are defined by the relation:

$$c_{n,k}(t) = \langle\varphi_{n,k}|\Psi(t)\rangle \qquad [3.46]$$

Let us apply the bra $\langle\varphi_{n,k}|$ to Schrödinger's equation [1.20]. We obtain:

$$i\hbar\frac{d\langle\varphi_{n,k}|\Psi(t)\rangle}{dt} = \langle\varphi_{n,k}|H|\Psi(t)\rangle = \langle\varphi_{n,k}|E_n|\Psi(t)\rangle \qquad [3.47]$$

Using [3.46], equation [3.47] is written as:

$$i\hbar\frac{dc_{n,k}(t)}{dt} = E_n c_{n,k}(t)$$

Hence:

$$\frac{dc_{n,k}}{c_{n,k}} = \frac{E_n}{i\hbar}dt$$

By integration, we find:

$$c_{n,k}(t) = c_{n,k}(t_0)e^{\frac{E_n}{i\hbar}(t-t_0)} \quad [3.48]$$

The state vectors at instants t_0 and t are then written as, respectively:

$$|\Psi(t_0)\rangle = \sum_{n,k}^{\infty} c_{n,k}(t_0)|\varphi_{n,k}\rangle \quad [3.49]$$

$$|\Psi(t)\rangle = \sum_{n,k}^{\infty} c_{n,k}(t)|\varphi_{n,k}\rangle = \sum_{n,k}^{\infty} c_{n,k}(t_0)e^{\frac{E_n}{i\hbar}(t-t_0)}|\varphi_{n,k}\rangle \quad [3.50]$$

In the particular case of the conservative systems described by stationary states, the initial ket $|\Psi(t_0)\rangle$ is itself eigenstate of the Hamiltonian H. Consequently, the expansion of ket $|\Psi(t_0)\rangle$ involves only the eigenstates of H associated with the same eigenvalues E_n. Summing is then done over index k, hence:

$$|\Psi(t_0)\rangle = \sum_{k}^{\infty} c_{n,k}(t_0)|\varphi_{n,k}\rangle \quad [3.51]$$

Taking [6.51] into account, the state vector [3.50] at instant t is written as:

$$|\Psi(t)\rangle = e^{\frac{E_n}{i\hbar}(t-t_0)}\sum_{n,k}^{\infty} c_{n,k}(t_0)|\varphi_{n,k}\rangle = e^{\frac{E_n}{i\hbar}(t-t_0)}|\Psi(t_0)\rangle \quad [3.52]$$

Equation [3.52] can be put in the form:

$$|\Psi(t)\rangle = e^{\frac{H}{i\hbar}(t-t_0)}|\Psi(t_0)\rangle = e^{\frac{E_n}{i\hbar}(t-t_0)}|\Psi(t_0)\rangle \quad [3.53]$$

The *evolution operator* $U(t, t_0)$ [1.34] has already been introduced. Equation [3.53] is then written as:

$$|\Psi(t)\rangle = U(t, t_0)|\Psi(t_0)\rangle \quad [3.54]$$

A comparison between [3.53] and [3.54] leads to the expression of the evolution operator [1.34] which can be rewritten as follows:

$$U(t,t_0)=e^{\frac{H}{i\hbar}(t-t_0)} \qquad [3.55]$$

3.3.3. *Ehrenfest's theorem*

The Hamiltonian for a conservative system is written as:

$$H=\frac{\vec{P}^2}{2m}+V(\vec{R}) \qquad [3.56]$$

Let us study the particular case of the evolution of the respective mean values $\langle\vec{R}\rangle$ and $\langle\vec{P}\rangle$ of the observables position $\vec{R}$ and linear momentum $\vec{P}$. Using [3.40], we get:

$$\frac{d\langle\vec{R}\rangle}{dt}=\frac{1}{i\hbar}\langle[\vec{R},H]\rangle+\left\langle\frac{\partial\vec{R}(t)}{\partial t}\right\rangle=\frac{1}{i\hbar}\langle[\vec{R},H]\rangle$$

$$\frac{d\langle\vec{P}\rangle}{dt}=\frac{1}{i\hbar}\langle[\vec{P},H]\rangle+\left\langle\frac{\partial\vec{P}(t)}{\partial t}\right\rangle=\frac{1}{i\hbar}\langle[\vec{P},H]\rangle$$

Using the expression [3.56] of the Hamiltonian, we obtain:

$$\frac{d\langle\vec{R}\rangle}{dt}=\frac{1}{i\hbar}\left\langle[\vec{R},\frac{\vec{P}^2}{2m}]\right\rangle+\frac{1}{i\hbar}\langle[R,V(\vec{R})]\rangle \qquad [3.57]$$

$$\frac{d\langle\vec{P}\rangle}{dt}==\frac{1}{i\hbar}\left\langle[\vec{P},\frac{\vec{P}^2}{2m}]\right\rangle+\frac{1}{i\hbar}\langle[\vec{P},V(\vec{R})]\rangle \qquad [3.58]$$

As already noted, an observable A commutes with its function: $[A, F(A)] = 0$. Moreover, according to [2.60], $[\vec{R},\vec{P}^2]=2i\hbar\vec{P}$.

$$\frac{d\langle\vec{R}\rangle}{dt}=\frac{1}{i\hbar}.\frac{2i\hbar\langle\vec{P}\rangle}{2m}$$

This yields:

$$\frac{d\langle\vec{R}\rangle}{dt}=\frac{\langle\vec{P}\rangle}{m} \qquad [3.59]$$

Similarly, taking [2.67] into account, we have:

$$\frac{d\langle\vec{P}\rangle}{dt}=\frac{1}{i\hbar}.(-i\hbar)\left\langle\frac{dG(\vec{R})}{dR}\right\rangle=-\langle\vec{\nabla}G(\vec{R})\rangle \qquad [3.60]$$

Summarizing results [3.59] and [3.60], we have:

$$\begin{cases}\dfrac{d\langle\vec{R}\rangle}{dt}=\dfrac{\langle\vec{P}\rangle}{m}\\[2ex] \dfrac{d\langle\vec{P}\rangle}{dt}=-\langle\vec{\nabla}V(\vec{R})\rangle\end{cases} \qquad [3.61]$$

The equations [3.61] of the evolution of the respective mean values of the observables position and linear momentum express what is known as *Ehrenfest's theorem*. Let us find the classical equivalent of these equations to make the connection with classical mechanics.

We consider the derivatives with respect to time of the position vector and linear momentum vector. Hence:

$$\begin{cases}\dfrac{d\overrightarrow{OM}}{dt}=\dfrac{d\vec{r}}{dt}=\vec{v}=\dfrac{\vec{p}}{m}\\[2ex] \vec{F}=\dfrac{d\vec{p}}{dt}=-\overrightarrow{grad}\,V(r)=-\vec{\nabla}V(r)\end{cases} \qquad [3.62]$$

A comparison between the systems of equations [3.61] and [3.62] reveals that Ehrenfest's theorem is the quantum equivalent of the derivative with respect to time of the position vector and of the fundamental law of dynamics.

Paul Ehrenfest was an Austrian theoretical physicist. In 1899, he studied at the Technische Hochschule of Vienna, where he attended **Boltzmann's lectures** (see Box A.3, Volume 1) on the "mechanical theory of heat". In 1904, under Boltzmann's supervision, in Vienna he defended his thesis on "The motion of rigid bodies in fluids and the mechanics of Hertz". Ehrenfest was the successor of **Lorentz** (see Box 3.9, Volume 1) at the chair of theoretical physics at the University of Leyde. In quantum mechanics, he is especially well known for the theorem bearing his name and reflecting the equations of evolution of the mean values of the observables position and linear momentum of a conservative system.

Box 3.1. *Ehrenfest (1880–1933)*

3.4. Exercises

3.4.1. *Exercise 1 – Pauli matrices, eigenvalues and eigenvectors*

Let us consider an orthonormal basis constituted by the eigenvectors $|+\rangle$ and $|-\rangle$ of the S_z observable. S_z is the component of the kinetic spin momentum operator according to the Oz direction. The expressions [2.110] of the Pauli matrix are recalled below:

$$\sigma_x = \begin{pmatrix} 0 & 1 \\ 1 & 0 \end{pmatrix} \sigma_y = \begin{pmatrix} 0 & -i \\ i & 0 \end{pmatrix} \sigma_z = \begin{pmatrix} 1 & 0 \\ 0 & -1 \end{pmatrix}$$

It is also worth recalling the spin angular momentum operator:

$$\vec{S} = \frac{\hbar}{2}\vec{\sigma}$$

(1) Determine the eigenvalues of σ_x, σ_y and σ_z. Deduce from them the eigenvalues of operators S_x, S_y and S_z.

(2) Determine the eigenvectors of σ_x, σ_y and σ_z. Deduce from them the eigenvectors of operators S_x and S_y.

3.4.2. *Exercise 2 – Observables associated with the spin*

Let us consider the two-dimensional spin states space subtended by the eigenvectors $|+\rangle$ and $|-\rangle$ of the spin observable S_z.

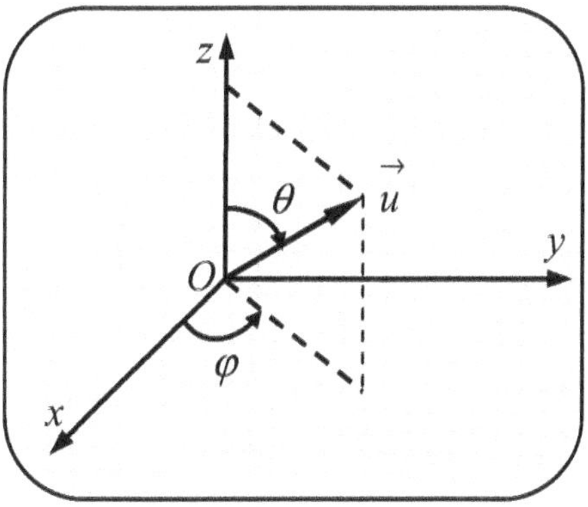

Figure 3.3. *Unit vector $\vec{u}$ defined by the polar coordinates θ and φ*

Furthermore, let l_u be the component of $\vec{l}$ along a unit vector $\vec{u}$ defined by the polar coordinates θ and φ (Figure 3.3).

The components l_x, l_y and l_z of the spin-related angular momentum $\vec{l}$ are associated, respectively, with the spin observables S_x, S_y and S_z represented in the basis $|+\rangle$ and $|-\rangle$ by the square matrices:

$$S_x = \frac{\hbar}{2}\begin{pmatrix} 0 & 1 \\ 1 & 0 \end{pmatrix} \quad S_y = \frac{\hbar}{2}\begin{pmatrix} 0 & -i \\ i & 0 \end{pmatrix} \quad S_z = \frac{\hbar}{2}\begin{pmatrix} 1 & 0 \\ 0 & -1 \end{pmatrix}$$

(1) Write the orthonormalization and closing equations satisfied by the vectors $|+\rangle$ and $|-\rangle$.

(2) Express l_u in the basis $|+\rangle$ and $|-\rangle$ as a function of l_x, l_y, l_z, θ and φ. Then deduce the matrix representing the observable S_u associated with l_u in the very same basis.

(3) Determine the eigenvectors $|+\rangle_u$ and $|-\rangle_u$ of the observable S_u.

For all practical purposes, the following trigonometric transformations are given:

$$(\cos x + 1) = 2\cos^2\left(\frac{x}{2}\right); \quad (\cos x - 1) = -2\sin^2\left(\frac{x}{2}\right)$$

3.4.3. *Exercise 3 – Evolution of a 1/2 spin in a magnetic field: CSCO, Larmor precession*

Let us consider a silver atom subjected to a uniform magnetic field $\vec{B}$ along direction Oz. The silver atom is in ground state.

(1) Express the potential energy of interaction W of the magnetic moment of the silver atom with the magnetic field. Then deduce the expression of the Hamiltonian H associated with W as a function of $\omega = 2\Omega$ (Ω: Larmor frequency) and of the spin observable to be specified.

(2) Prove that H and S_z constitute a CSCO.

(3) Then determine the eigenvalues E^+ and E^- of H associated, respectively, with the eigenvectors $|+\rangle$ and $|-\rangle$.

(4) The purpose is now to evidence the Larmor precession. For this, we consider the evolution in the magnetic field of the spin magnetic moment $\vec{M} = M_u \vec{u}$, where $\vec{u}$ is the unit vector identifying the direction of $\vec{M}$ in polar coordinates (Figure 3.4).

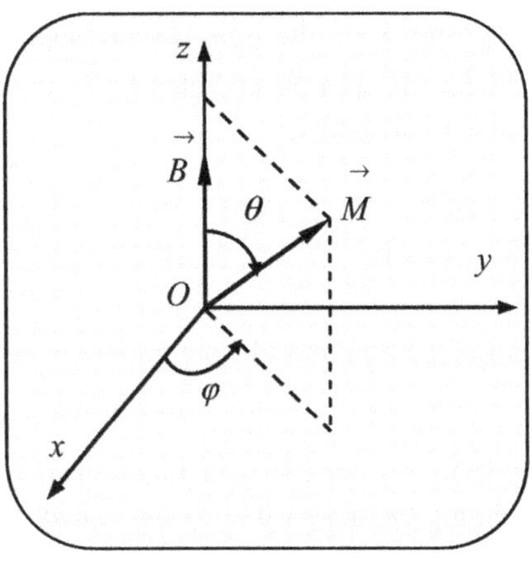

Figure 3.4. *Spin magnetic moment $\vec{M} = M_u\vec{u}$ in polar coordinates*

We assume that at the initial instant $t_0 = 0$, the eigenstate of the spin associated with the eigenvalue $+\hbar/2$ is given by the ket vector:

$$|\Psi(t_0 = 0)\rangle = \cos\frac{\theta}{2} e^{-i\varphi/2}|+\rangle + \sin\frac{\theta}{2} e^{i\varphi/2}|-\rangle$$

(4.1) Express the expansion of the state vector $|\Psi(t)\rangle$ over the eigenstates of Hamiltonian H.

(4.2) Prove that $|\Psi(t)\rangle$ can be written in the following form:

$$|\Psi(t)\rangle = \cos\frac{\theta}{2} e^{-i\frac{\phi(t)}{2}} |+\rangle + \sin\frac{\theta}{2} e^{i\frac{\phi(t)}{2}} |-\rangle$$

Draw a conclusion.

(4.3) Specify the polar coordinates of the unit vector $\vec{u}(t)$. Prove that the spin magnetic moment is in Larmor precession motion.

3.4.4. *Exercise 4 – Eigenvalue of the squared angular momentum operator*

Let l_x, l_y and l_z be the operators of angular momentum operator projections on the axes of coordinates x, y and z. Let $\vec{l}^2$ be the square angular momentum operator. Moreover, we introduce the operators denoted by l_+ and l_- and defined by the expressions:

$$l_+ = l_x + il_y;\ l_- = l_x - il_y$$

(1) Prove that the square $\vec{l}^2$ of the angular momentum operator commutes with one of its components, for example l_z. Draw a conclusion.

(2) Calculate the commutators $[l_+, l_-]$, $[l_z, l_+]$ and $[l_z, l_-]$.

(3) Express $\vec{l}^2$ as a function of its components. Prove the relations:

$$\vec{l}^2 = l_+ l_- + l_z^2 - \hbar l_z = l_- l_+ + l_z^2 + \hbar l_z$$

(4) Let λ^2 be the eigenvalue of l^2. Moreover, $|\Phi\rangle$ designates the eigenstate common to l^2 and l_z. Write the equations with eigenvalues of operators l^2 and l_z.

(5) Then prove that the states $l_+|\Phi\rangle$ and $l_-|\Phi\rangle$ are eigenstates of l_z associated, respectively, with the eigenvalues α_+ and α_- to be specified.

(6) Which one of the eigenvalues α_+ or α_- is incompatible with the principle of quantization of operator l_z for its maximal eigenvalue? Justify the answer.

(7) Then justify the fact that we must put $l_+|\Phi\rangle = 0$. Deduce the expression of the eigenvalue λ^2 of the square angular momentum operator $\vec{l}^{\,2}$.

(8) Draw a vector diagram to represent the possible orientations of the angular momentum operator in unit $\hbar$ when $\ell = 2$.

3.4.5. *Exercise 5 – Constant of motion, good quantum numbers*

Let us consider an observable A that is explicitly time independent and commutes with the Hamiltonian H, which is itself explicitly time independent. Let $\{|\Phi_{n,k,m}\rangle\}$ be a system of eigenkets common to A and H, the discrete index m identifying the eigenvalues of observables forming a CSCO with H and A. The eigenvalues of H and A are denoted by E_n and a_k, respectively.

(1) Write the equations with eigenvalues of observables H and A.

(2) Prove that H and A are constants of motion.

(3) The system is assumed in state $|\Phi_{n,k,m}\rangle$ at the initial instant t_0. Explain why the eigenvalues of A are known as *good quantum numbers*.

(4) Express the expansions of state vectors $|\Psi\ (t_0)\rangle$ and $|\Psi\ (t)\rangle$ over the eigenstates $|\Phi_{n,k,m}\rangle$ common to observables H and A.

3.4.6. *Exercise 6 – Evolution of the mean values of the operators associated with position and linear momentum*

We consider the operators X and P (for P_x) associated, respectively, with the position x and the linear momentum p of a free particle.

(1) Using Ehrenfest's theorem, establish the equations of evolution of the mean values $\langle X\rangle$ and $\langle P\rangle$. Make the connection with classical mechanics.

(2) Establish also the equations of evolution of the mean values $\langle X^2\rangle$, $\langle P^2\rangle$ and $\langle XP + PX\rangle$. Make the connection with classical mechanics.

(3) Express the quadratic deviation ΔX. What happens when t increases? Draw a conclusion. We consider that at $t = 0$, $\langle XP + PX\rangle_0 = \langle X\rangle_0 = 0$.

3.4.7. *Exercise 7 – Particle subjected to various potentials*

A particle of mass m is immersed in a field of forces whose potential is $V(x) = a\,x^n$, where a is a constant and n is an integer.

(1) Establish the equations of evolution of $\langle X \rangle$ and $\langle P \rangle$ when the particle is:

(1.1) Free.

(1.2) Subjected to a uniform potential.

(1.3) Subjected to a parabolic potential.

(2) Then make the connection with classical mechanics.

(3) Does the particle behave as a classical system when $n = 3$? Draw a conclusion in the general case.

3.4.8. *Exercise 8 – Oscillating molecular dipole, root mean square deviation*

We reconsider the model of the hydrogen chloride molecule assimilated to an oscillating dipole such as that described in the exercise in section 1.7.12 in Chapter 1 (Volume 2).

The lighter hydrogen atom is maintained at a mean distance a from the center of inertia of chlorine. The potential energy thus constituted is:

$$V(x) = \frac{1}{2}k(x-a)^2$$

Moreover, the normed wave functions $\Phi_0(x)$ of the fundamental state and $\Phi_1(x)$ of the first excited state are given by the following expressions:

$$\Phi_0(x) = \left(\frac{m\omega}{\pi\hbar}\right)^{\frac{1}{4}} e^{-\frac{1}{2}\frac{m\omega}{\hbar}(x-a)^2}$$

$$\Phi_1(x) = \left[\frac{4}{\pi}\left(\frac{m\omega}{\hbar}\right)\right]^{\frac{1}{4}} (x-a)\, e^{-\frac{1}{2}\frac{m\omega}{\hbar}(x-a)^2}$$

(1) Calculate the mean value $\langle X\rangle$ of the observable X when the oscillator is in ground state and when it is in the first excited state.

(2) The uncertainty involved in the measurement of position of the hydrogen atom relative to that of the chlorine atom is defined by the relation:

$$\Delta X = \sqrt{\langle\Phi|(X-a)^2|\Phi\rangle}$$

(2.1) Calculate the uncertainties $(\Delta X)_0$ and $(\Delta X)_1$ when the oscillating system is in ground state and in the first excited state.

(2.2) Provide an interpretation of the observed difference between $(\Delta X)_0$ and $(\Delta X)_1$.

Given data. For the family of integrals of the type:

$$I_p = \int_0^\infty x^p e^{-\rho x^2} dx \ (\rho > 0),$$

the recurrence relation is written as:

$$I_p = \frac{p-1}{2\rho} I_{p-2}, \text{ with } I_0 = \frac{1}{2}\sqrt{\frac{\pi}{\rho}} \text{ and } I_1 = \frac{1}{2\rho}$$

3.4.9. *Exercise 9 – Infinite potential well, time–energy uncertainty relation*

Let us consider a particle confined in an infinitely deep potential well of width l such that:

$$V(x) = 0, \text{ if } 0 \le x \le l$$

$$V(x) = \infty, \text{ elsewhere.}$$

The Hamiltonian of the particle is:

$$H = \frac{P^2}{2m} + V(X)$$

The eigenvalues of the Hamiltonian H are quantized and given by the expression (see formula [1.62]):

$$E_n = \frac{\hbar^2 \pi^2}{2ml^2} n^2$$

Moreover, the eigenfunctions of H are given by (see formula [1.67]):

$$\Phi_n(x) = \sqrt{\frac{2}{l}} \sin\left(\frac{n\pi x}{l}\right), \text{ if } 0 < x < l$$

$\Phi_n(x) = 0$, elsewhere.

Wave functions $\Phi_n(x)$ are associated with kets $|\Phi_n\rangle$.

At $t = 0$, the state of the particle is described by the ket $|\Psi(0)\rangle$ whose expansion in the basis $\{|\Phi_n\rangle\}$ is written as:

$$|\Psi(0)\rangle = \frac{1}{\sqrt{2}}\left(|\Phi_1\rangle + |\Phi_2\rangle\right)$$

(1) Express the wave function $\Psi(x, 0)$ in $\{|x\rangle\}$ representation.

(2) Prove that at instant t, the state vector $|\Psi(t)\rangle$ is written as:

$$|\Psi(t)\rangle = \frac{1}{\sqrt{2}}\left(e^{-i\alpha_1 t}|\Phi_1\rangle + e^{-i\alpha_2 t}|\Phi_2\rangle\right)$$

In this relation, α_1 and α_2 are constants to be specified.

(3) Prove that the density of the probability of presence $D(x, t)$ of the particle can be put in the form:

$$D(x,t) = \frac{1}{2}\left(\Phi_1^2(x) + \Phi_2^2(x) + 2\Phi_1(x)\Phi_2(x)\cos\omega_{21}t\right)$$

In this expression, ω_{21} is a Bohr angular frequency to be specified. Draw a conclusion.

(4) Calculate the mean value $\langle X\rangle(t)$ representing the motion of the center of the wave packet associated with the particle. We put $X' = X - l/2$.

(5) Represent the variations of the mean value $\langle X\rangle(t)$ as a function of time compared to the predictions of classical mechanics (*a classical particle that is moving back and forth in a well with angular frequency* ω_{21} is considered).

(6) Calculate $\langle H\rangle(t)$ and $\langle H^2\rangle(t)$. Deduce the root mean square deviation ΔH. Are these quantities time-dependent? Explain why.

(7) Knowing that the wave packet significantly evolves after a period of time of about $1/\omega_{21}$, find the time–energy uncertainty relation.

Given data.

$$\int_0^a \left(q-\frac{a}{2}\right)\sin^2\left(\frac{n\pi q}{a}\right)dq = 0$$

$$\int_0^a q\sin\left(\frac{\pi q}{a}\right)\sin\left(\frac{2\pi q}{a}\right)dq = -\frac{16\,a^2}{18\,\pi^2}$$

3.4.10. *Exercise 10 – Study of a conservative system*

Let us consider a conservative system whose Hamiltonian is H_0. The space of states of the system is generated by the basis vectors $|\Phi_n\rangle$, $n = 1, 2, 3$. Let E_0 be the eigenvalue of H_0 such that:

$$H_0|\Phi_n\rangle = E_0|\Phi_n\rangle$$

Moreover, an operator W generating a coupling between various vectors $|\Phi_n\rangle$ is considered, with:

$$\langle\Phi_1|W|\Phi_3\rangle = \langle\Phi_3|W|\Phi_1\rangle = \langle\Phi_1|W|\Phi_2\rangle = \langle\Phi_2|W|\Phi_1\rangle = \frac{\lambda}{\sqrt{2}}$$

where λ is a constant such that $0 < \lambda < E_0$.

(1) Write the matrix representing the Hamiltonian $H = H_0 + W$ in the discrete set $\{|\Phi_n\rangle\}$.

(2) We designate by $|\phi_-\rangle$, $|\phi_0\rangle$ and $|\phi_+\rangle$ the eigenstates of H associated, respectively, with eigenvalues α_-, α_0 and α_+. Express:

(2.1) The eigenvalues α_-, α_0 and α_+ as a function of λ and E_0.

(2.2) The eigenstates $|\phi_-\rangle$, $|\phi_0\rangle$ and $|\phi_+\rangle$ of H.

(3) At instant $t = 0$, the system is in the state $|\Psi(0)\rangle = |\Phi_1\rangle$.

(3.1) Find the state $|\Psi(t)\rangle$ of the system at instant t.

(3.2) Calculate the mean value $\langle H\rangle(t)$. Draw a conclusion.

3.4.11. *Exercise 11 – Evolution of the density operator*

Let us consider a *statistical mixture of states* of a system. In the context of quantum mechanics, it can be proven that it is not possible to describe the mixture by a "mean vector" [COH 77]. The introduction of the density operator facilitates a simple description of the statistical mixture of states.

In this exercise, we consider the case of a system in a *pure state* (the state of the system is perfectly determined). Let $\{|u_n\rangle\}$ be a discrete orthonormal basis in the space of states of the system. At the instant t, a normed ket vector of the system can be written as:

$$|\Psi(t)\rangle = \sum_n c_n(t)|u_n\rangle$$

Moreover, the density operator is defined in $\{|u_n\rangle\}$ representation as:

$$\rho(t) = |\Psi(t)\rangle\langle\Psi(t)|$$

Let A be an observable.

(1) Find the trace of $\rho(t)$ and then express $\langle A\rangle$ as a function of $\rho(t)$ and A.

(2) Find the equation of evolution followed by $\rho(t)$.

(3) We designate by $|u_n\rangle$ the eigenkets of the Hamiltonian H associated with the eigenvalues E_n. Prove the relation:

$$i\hbar\frac{d}{dt}\rho_{n,k}(t) = (E_n - E_k)\cdot\rho_{n,k}(t)$$

(4) Prove that the *populations* $\rho_{nn}(t)$ are constant and that *coherences* $\rho_{nk}(t)$ oscillate at Bohr's frequency, ν_{nk}, to be specified.

3.4.12. *Exercise 12 – Evolution of a spin 1/2 in a magnetic field*

Let us consider a silver atom subjected to a uniform magnetic field $\vec{B}$ along direction Oz. The silver atom is in ground state. The Hamiltonian H describing the spin evolution in the magnetic field is given by the relation:

$$H = \omega S_z$$

In this relation, S_z is the observable associated with the spin along direction Oz and whose eigenstates are $|+\rangle$ and $|-\rangle$, $\omega = -2\gamma B$, where γ is the gyromagnetic ratio of the electron. Moreover, using the equation with eigenvalues of the Hamiltonian, we get:

$$\begin{cases} H|+\rangle = E^+|+\rangle = \dfrac{\hbar\omega}{2}|+\rangle \\ H|-\rangle = E^-|-\rangle = -\dfrac{\hbar\omega}{2}|-\rangle \end{cases}$$

We assume that the spin at $t = 0$ is in the state:

$$|\Psi(0)\rangle = \cos\frac{\theta}{2}e^{-i\varphi/2}|+\rangle + \sin\frac{\theta}{2}e^{i\varphi/2}|-\rangle$$

(1) Verify that the physical quantities associated with observables H and S_z are simultaneously measurable.

(2) Calculate the mean values $\langle H\rangle_0$ and $\langle S_z\rangle_0$ at $t = 0$. Draw a conclusion.

(3) The system is now considered at a given instant t.

(3.1) Express the state vector at instant t.

(3.2) Find the mean values $\langle H\rangle (t)$ and $\langle S_z\rangle (t)$. Draw a conclusion.

(3.3) Is it possible to simultaneously measure the energy E and the components associated with S_x and S_y of the spin angular momentum? Justify the answer.

(3.4) Calculate the mean values $\langle S_x\rangle (t)$ and $\langle S_y\rangle (t)$. Draw a conclusion.

Given data. The matrices representing the observables S_x and S_y in the basis constituted of kets $|+\rangle$ and $|-\rangle$:

$$(S_x) = \frac{\hbar}{2}\begin{pmatrix} 0 & 1 \\ 1 & 0 \end{pmatrix}; \ (S_y) = \frac{\hbar}{2}\begin{pmatrix} 0 & -i \\ i & 0 \end{pmatrix}$$

3.5. Solutions

3.5.1. *Solution 1 – Pauli matrices, eigenvalues and eigenvectors*

(1) *Finding the eigenvalues*

Let us consider the characteristic equation Det $[A - \lambda I] = 0$ and then evaluate the matrices $\sigma_x - \lambda I$, $\sigma_y - \lambda I$ and $\sigma_z - \lambda I$. Using Pauli matrices [2.110], we find:

$$\sigma_x - \lambda I = \begin{pmatrix} 0 & 1 \\ 1 & 0 \end{pmatrix} - \lambda \begin{pmatrix} 1 & 0 \\ 0 & 1 \end{pmatrix} = \begin{pmatrix} -\lambda & 1 \\ 1 & -\lambda \end{pmatrix}$$

$$\sigma_y - \lambda I = \begin{pmatrix} 0 & -i \\ i & 0 \end{pmatrix} - \lambda \begin{pmatrix} 1 & 0 \\ 0 & 1 \end{pmatrix} = \begin{pmatrix} -\lambda & -i \\ i & -\lambda \end{pmatrix}$$

$$\sigma_z - \lambda I = \begin{pmatrix} 1 & 0 \\ 0 & -1 \end{pmatrix} - \lambda \begin{pmatrix} 1 & 0 \\ 0 & 1 \end{pmatrix} = \begin{pmatrix} 1-\lambda & 0 \\ 0 & -1-\lambda \end{pmatrix}$$

These expressions show that the three matrices have the same secular equation:

$$\lambda^2 - 1 = 0 \Rightarrow \lambda = \pm 1 \qquad [3.63]$$

The eigenvalues of operators σ_x, σ_y and σ_z are therefore $\lambda = \pm 1$.

The spin angular momentum operator is given by the relation:

$$\vec{S} = \frac{\hbar}{2}\vec{\sigma} \qquad [3.64]$$

Using [3.63], we deduce the eigenvalues of operators S_x, S_y and S_z:

$$\lambda' = \pm \frac{\hbar}{2} \qquad [3.65]$$

(2) *Finding the eigenvectors*

Relation [3.64] shows that operators σ_z and S_z have the same eigenvectors $|+\rangle$ and $|-\rangle$. The equations with eigenvalues of σ_z are thus written as:

$$\begin{cases} \sigma_z |+\rangle = +|+\rangle \\ \sigma_z |-\rangle = -|-\rangle \end{cases} \qquad [3.66]$$

Let us designate by $|\pm\rangle_x$ and $|\pm\rangle_y$, respectively, the eigenvectors of operators σ_x and σ_y. We express their expansions on the basis vectors of operator S_z. The equations with eigenvalues of σ_x and σ_y are written as follows:

$$\begin{cases}\sigma_x|+\rangle_x = +|+\rangle_x \\ \sigma_x|-\rangle_x = -|-\rangle_x\end{cases} \Rightarrow \begin{cases}\sigma_x|\pm\rangle_x = \pm|\pm\rangle_x \\ \sigma_y|\pm\rangle_y = \pm|\pm\rangle_y\end{cases} \quad \begin{cases}\sigma_y|+\rangle_y = +|+\rangle_y \\ \sigma_y|-\rangle_y = -|-\rangle_y\end{cases} \tag{3.67}$$

Let $\alpha_\pm$ and $\beta_\pm$ be the coefficients of the expansions of the eigenvectors of σ_x and σ_y. We have:

$$\begin{aligned}|+\rangle_x &= \alpha_+|+\rangle + \beta_+|-\rangle \\ |-\rangle_x &= \alpha_-|+\rangle + \beta_-|-\rangle\end{aligned} \tag{3.68}$$

– *Eigenvectors of* σ_x

According to [3.68], the vectors $|\pm\rangle_x$ are represented in the basis of eigenstates of S_z by the single column matrices:

$$|+\rangle_x = \begin{pmatrix}\alpha_+ \\ \beta_+\end{pmatrix}, |-\rangle_x = \begin{pmatrix}\alpha_- \\ \beta_-\end{pmatrix} \tag{3.69}$$

Using [3.69] and Pauli matrices, we get, according to [3.67]:

$$\begin{pmatrix}0 & 1 \\ 1 & 0\end{pmatrix}\begin{pmatrix}\alpha_\pm \\ \beta_\pm\end{pmatrix} = \pm\begin{pmatrix}\alpha_\pm \\ \beta_\pm\end{pmatrix} \Rightarrow \begin{cases}\alpha_+ = +\beta_+ \\ \alpha_- = -\beta_-\end{cases} \tag{3.70}$$

Moreover, the kets $|\pm\rangle_x$ being normed, according to the normalization condition we have:

$$|\alpha_\pm|^2 + |\beta_\pm|^2 = 1 \Rightarrow 2|\alpha_\pm|^2 = 2|\beta_\pm|^2 = 1$$

If coefficients $\alpha_\pm$ and $\beta_\pm$ are arbitrarily chosen real, we get:

$$\begin{cases}\alpha_+ = +\beta_+ = \dfrac{1}{\sqrt{2}} \\ \alpha_- = -\beta_- = \dfrac{1}{\sqrt{2}}\end{cases} \tag{3.71}$$

Using [3.71], the eigenvectors of operator σ_x are written according to [3.67]:

$$\begin{cases} |+\rangle_x = \frac{1}{\sqrt{2}}\left[\,|+\rangle + |-\rangle\,\right] \\ |-\rangle_x = \frac{1}{\sqrt{2}}\left[\,|+\rangle - |-\rangle\,\right] \end{cases} \qquad [3.72]$$

– *Eigenvectors of* σ_y

According to the above, vectors $|\pm\rangle_y$ are represented in the basis of states of S_z by the same matrices [3.69] (a change in coefficients is not required). Using Pauli matrices [2.110], we get according to [3.67]:

$$\begin{pmatrix} 0 & -i \\ i & 0 \end{pmatrix}\begin{pmatrix} \alpha_\pm \\ \beta_\pm \end{pmatrix} = \pm\begin{pmatrix} \alpha_\pm \\ \beta_\pm \end{pmatrix}$$

This leads to:

$$\begin{cases} -i\beta_+ = +\alpha_+ \\ i\alpha_+ = +\beta_+ \end{cases} ; \begin{cases} -i\beta_- = -\alpha_- \\ i\alpha_- = -\beta_- \end{cases} \qquad [3.73]$$

Considering [3.73] and taking advantage of the normalization condition, we have:

$$\begin{cases} \alpha_+ = -i\,\beta_+ \\ \alpha_- = i\,\beta_- \end{cases} \Rightarrow \begin{cases} \alpha_+ = -i\beta_+ = \frac{1}{\sqrt{2}} \\ \alpha_- = i\beta_- = \frac{1}{\sqrt{2}} \end{cases} \qquad [3.74]$$

Inserting results [3.74] into equation [3.68] where index x should be replaced by y, we find the expressions of eigenvectors of the operator σ_y. Hence:

$$\begin{cases} |+\rangle_y = \frac{1}{\sqrt{2}}\left[\,|+\rangle + i|-\rangle\,\right] \\ |-\rangle_y = \frac{1}{\sqrt{2}}\left[\,|+\rangle - i|-\rangle\,\right] \end{cases} \qquad [3.75]$$

According to [3.64], it can be noted that operators S_x and S_y have, respectively, the same eigenvectors [3.72] and [3.75] as operators σ_x and σ_y.

3.5.2. *Solution 2 – Observables associated with the spin*

The matrices representing the spin observables S_x, S_y and S_z in the basis $|+\rangle$ and $|-\rangle$ are:

$$S_x = \frac{\hbar}{2}\begin{pmatrix} 0 & 1 \\ 1 & 0 \end{pmatrix};\ S_y = \frac{\hbar}{2}\begin{pmatrix} 0 & -i \\ i & 0 \end{pmatrix};\ S_z = \frac{\hbar}{2}\begin{pmatrix} 1 & 0 \\ 0 & -1 \end{pmatrix} \qquad [3.76]$$

(1) *Orthonormalization and closing equations*

The orthonormalization and closing relations satisfied by the eigenvectors $|+\rangle$ and $|-\rangle$ are the following:

– Orthonormalization relation

$$\begin{cases} \langle +|+\rangle = \langle -|-\rangle = 1 \\ \langle +|-\rangle = \langle -|+\rangle = 0 \end{cases} \qquad [3.77]$$

– Closing relation

$$|+\rangle\langle +| + |-\rangle\langle -| = \mathbb{1} \qquad [3.78]$$

(2) *Expression, matrix of the observable S_u*

Let us consider Figure 3.5.

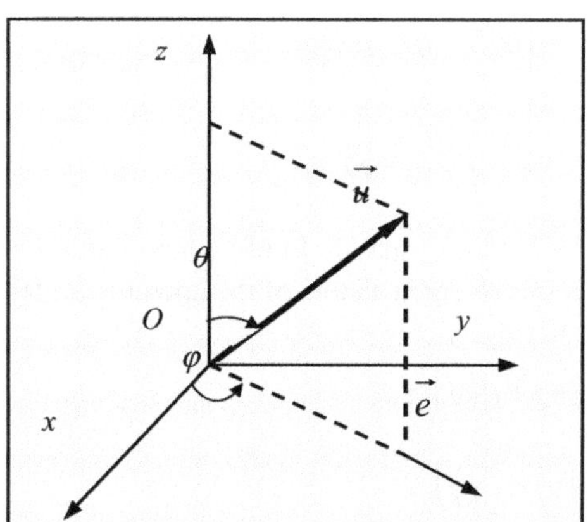

Figure 3.5. *Orthonormal basis $\{\vec{e}, \vec{k}\}$*

In the $\{\vec{e}, \vec{k}\}$ basis, the unit vector $\vec{u}$ is expressed by the following relation:

$$\vec{u} = (\vec{u}.\vec{e}).\vec{e} + (\vec{u}.\vec{z}).\vec{z} \Rightarrow \vec{u} = \sin\theta.\vec{e} + \cos\theta.\vec{z} \qquad [3.79]$$

But the unit vector $\vec{e}$ is written as:

$$\vec{e} = \cos\varphi.\vec{x} + \sin\varphi.\vec{y}$$

The unit vector [3.79] is then written as:

$$\vec{u} = \sin\theta\cos\varphi.\vec{x} + \sin\theta\sin\varphi.\vec{y} + \cos\theta.\vec{z} \qquad [3.80]$$

Moreover, l_u is written as:

$$l_u = \vec{l}.\vec{u}\text{, with } \vec{l} = l_x\vec{x} + l_y\vec{y} + l_z\vec{z}$$

Considering [3.80], we get:

$$l_u = l_x \sin\theta\cos\varphi + l_y \sin\theta\sin\varphi + l_z \cos\theta \qquad [3.81]$$

The expression of the observable S_u associated with l_u can be deduced from [3.81]. Hence:

$$S_u = S_x \sin\theta\cos\varphi + S_y \sin\theta\sin\varphi + S_z \cos\theta \qquad [3.82]$$

Using [3.76], we express the matrix representing the observable S_u in this same basis $\{|+\rangle, |-\rangle\}$. Hence:

$$S_u = \frac{\hbar}{2}\begin{pmatrix} \cos\theta & \sin\theta(\cos\varphi - i\sin\varphi) \\ \sin\theta(\cos\varphi + i\sin\varphi) & -\cos\theta \end{pmatrix}$$

Finally:

$$S_u = \frac{\hbar}{2}\begin{pmatrix} \cos\theta & \sin\theta e^{-i\varphi} \\ \sin\theta e^{i\varphi} & -\cos\theta \end{pmatrix} \qquad [3.83]$$

(3) *Eigenvectors*

The equations with eigenvalues of the observable S_u are:

$$\begin{cases} S_u|+\rangle_u = \frac{\hbar}{2}|+\rangle_u \\ S_u|-\rangle_u = -\frac{\hbar}{2}|-\rangle_u \end{cases} \qquad [3.84]$$

Let $\alpha_\pm$ and $\beta_\pm$ be the coefficients of the expansion of kets $|\pm\rangle_u$ on the basis $\{|+\rangle, |-\rangle\}$. We have:

$$\begin{cases} |+\rangle_u = \alpha_+|+\rangle + \beta_+|-\rangle \\ |-\rangle_u = \alpha_-|+\rangle + \beta_-|-\rangle \end{cases} \qquad [3.85]$$

According to [3.85], the kets $|\pm\rangle_u$ are represented in the basis $\{|+\rangle,|-\rangle\}$ by the matrices:

$$|+\rangle_u = \begin{pmatrix} \alpha_+ \\ \beta_+ \end{pmatrix}; |-\rangle_u = \begin{pmatrix} \alpha_- \\ \beta_- \end{pmatrix} \qquad [3.86]$$

Using [3.83] and [3.86], the equations with eigenvalues [3.84] are written as:

$$\begin{pmatrix} \cos\theta & \sin\theta e^{-i\varphi} \\ \sin\theta e^{i\varphi} & -\cos\theta \end{pmatrix} \begin{pmatrix} \alpha_\pm \\ \beta_\pm \end{pmatrix} = \pm \begin{pmatrix} \alpha_\pm \\ \beta_\pm \end{pmatrix} \qquad [3.87]$$

This equality leads to:

$$\begin{cases} \alpha_\pm \cos\theta + \beta_\pm \sin\theta e^{-i\varphi} = \pm\alpha_\pm \\ \alpha_\pm \sin\theta e^{i\varphi} - \beta_\pm \cos\theta = \pm\beta_\pm \end{cases} \qquad [3.88]$$

Let us solve the first system of equations [3.88] relative to coefficient α_+. We have:

$$\begin{cases} \alpha_+(\cos\theta - 1) + \beta_+ \sin\theta e^{-i\varphi} = 0 \\ \alpha_+ \sin\theta e^{i\varphi} - \beta_+(\cos\theta + 1) = 0 \end{cases} \qquad [3.89]$$

Using the following transformations:

$$(\cos\theta + 1) = 2\cos^2\left(\frac{\theta}{2}\right) ; (\cos\theta - 1) = -2\sin^2\left(\frac{\theta}{2}\right)$$

The system [3.89] is then written as:

$$\begin{cases} -\alpha_+ \sin\frac{\theta}{2} \quad + \beta_+ \cos\frac{\theta}{2} e^{-i\varphi} = 0 \\ \alpha_+ \sin\frac{\theta}{2} e^{i\varphi} \quad - \beta_+ \cos\frac{\theta}{2} \quad = 0 \end{cases} \quad [3.90]$$

These equations lead to:

$$\beta_+ = \alpha_+ \tan\frac{\theta}{2} e^{i\varphi} ; |\alpha_+|^2 + |\beta_+|^2 = 1 \quad [3.91]$$

The last relation [3.91] expresses the normalization condition verified by the ket $|+\rangle_u$. Then we obtain:

$$|\alpha_+|^2 \left(1 + \tan^2\frac{\theta}{2}\right) = 1 \Rightarrow |\alpha_+| = \cos\frac{\theta}{2} \quad [3.92]$$

Knowing that coefficients α_+ and β_+ depend on φ, we choose α_+ so that it is proportional to $e^{-i\varphi/2}$ (the factor $e^{-i\varphi}$ would yield a coefficient β_+ that does not depend on φ, as shown by the relation [3.91]). Hence:

$$\begin{cases} \alpha_+ = \cos\frac{\theta}{2} e^{-i\varphi/2} \\ \beta_+ = \sin\frac{\theta}{2} e^{i\varphi/2} \end{cases} \quad [3.93]$$

Using the second system of equations [3.89] relative to coefficient α_-, and adopting a similar reasoning as previously, we get:

$$\begin{cases} \alpha_- = -\sin\frac{\theta}{2} e^{-i\varphi/2} \\ \beta_- = \cos\frac{\theta}{2} e^{i\varphi/2} \end{cases} \quad [3.94]$$

Using [3.93] and [3.94], the eigenvectors [3.85] are written as:

$$\begin{cases} |+\rangle_u = \cos\frac{\theta}{2}e^{-i\varphi/2}|+\rangle + \sin\frac{\theta}{2}e^{i\varphi/2}|-\rangle \\ |-\rangle_u = -\sin\frac{\theta}{2}e^{-i\varphi/2}|+\rangle + \cos\frac{\theta}{2}e^{i\varphi/2}|-\rangle \end{cases} \quad [3.95]$$

3.5.3. *Solution 3 – Evolution of a 1/2 spin in a magnetic field: CSCO, Larmor precession*

(1) *Potential energy*

The potential energy W of interaction between the magnetic moment of the silver atom and the magnetic field is:

$$W = -\vec{M}\cdot\vec{B} = -g_S\gamma\vec{s}\cdot\vec{B} \Rightarrow W = -g_S\gamma B s_z \quad [3.96]$$

Inserting the Larmor frequency $\Omega = -\gamma B$ into the expression above and knowing the Landé factor $g_s = 2$, we get:

$$W = 2\Omega s_z \quad [3.97]$$

We consider $\omega = 2\Omega$. The observable associated with spin S_z, the Hamiltonian H associated with W, is then written as:

$$H = \omega S_z \quad [3.98]$$

(2) *CSCO*

Relation [3.98] expresses that observables H and S_z commute. Indeed, we have:

$$[H, S_z] = [\omega S_z, S_z] = \omega[S_z, S_z] = 0 \quad [3.99]$$

Moreover, H is proportional to S_z according to [3.98], and therefore they have the same eigenvectors (eigenvectors $|+\rangle$ and $|-\rangle$ of S_z): H and S_z therefore constitute a CSCO.

(3) *Eigenvalues*

Let us designate by E^+ and E^- the eigenvalues of H associated, respectively, with eigenvectors $|+\rangle$ and $|-\rangle$. The equation with eigenvalues is written as follows:

$$\begin{cases} H|+\rangle = \omega S_z|+\rangle = \frac{\hbar\omega}{2}|+\rangle \\ H|-\rangle = \omega S_z|-\rangle = -\frac{\hbar\omega}{2}|-\rangle \end{cases} \Rightarrow \begin{cases} H|+\rangle = E^+|+\rangle \\ H|-\rangle = E^-|-\rangle \end{cases} \qquad [3.100]$$

or:

$$E^+ = \frac{\hbar\omega}{2} ; E^- = -\frac{\hbar\omega}{2} \qquad [3.101]$$

Relations [3.101] show that in the magnetic field, the silver atom has two energy levels. The energy gap between the higher level E^+ (positive angular frequency ω) and the lower level E^- is equal to $\hbar\omega$.

(4) *Spin magnetic moment*

$$\overrightarrow{M} = M_u \vec{u}$$

The initial state of the spin associated with the eigenvalue $+\hbar/2$ is described by the ket:

$$|\Psi(0)\rangle = \cos\frac{\theta}{2} e^{-i\frac{\varphi}{2}}|+\rangle + \sin\frac{\theta}{2} e^{i\frac{\varphi}{2}}|-\rangle \qquad [3.102]$$

(4.1) *Expression of the state vector*

At the instant t, the evolution of ket $|\Psi(t)\rangle$ is governed by the equation:

$$|\Psi(t)\rangle = U(t,t_0)|\Psi(0)\rangle ; U(t,t_0) = e^{-iHt/\hbar} \qquad [3.103]$$

Using [3.102], the evolution equation [3.103] is written as:

$$|\Psi(t)\rangle = \cos\frac{\theta}{2} e^{-i\varphi/2} U(t,t_0)|+\rangle + \sin\frac{\theta}{2} e^{i\varphi/2} U(t,t_0)|-\rangle$$

Hence:

$$|\Psi(t)\rangle = \cos\frac{\theta}{2} e^{-i\varphi/2} e^{-iHt/\hbar}|+\rangle + \sin\frac{\theta}{2} e^{i\varphi/2} e^{-iHt/\hbar}|-\rangle$$

Or:

$$|\Psi(t)\rangle = \cos\frac{\theta}{2} e^{-i\varphi/2} e^{-iE^{+}t/\hbar}|+\rangle + \sin\frac{\theta}{2} e^{i\varphi/2} e^{-iE^{-}t/\hbar}|-\rangle \qquad [3.104]$$

Using [3.101], the evolution equation [3.104] can be written after arrangement:

$$|\Psi(t)\rangle = \cos\frac{\theta}{2} e^{-i\frac{(\omega t+\varphi)}{2}}|+\rangle + \sin\frac{\theta}{2} e^{i\frac{(\omega t+\varphi)}{2}}|-\rangle \qquad [3.105]$$

Putting $\phi(t) = \omega t + \varphi$, the ket [3.105] is finally written as:

$$|\Psi(t)\rangle = \cos\frac{\theta}{2} e^{-i\frac{\phi(t)}{2}}|+\rangle + \sin\frac{\theta}{2} e^{i\frac{\phi(t)}{2}}|-\rangle \qquad [3.106]$$

CONCLUSION.– Since $\omega = -2\gamma B$, then:

$$\phi(t) = -2\gamma Bt + \varphi \qquad [3.107]$$

Relation [3.107] shows that the magnetic field introduces a phase difference between the coefficients assigned to eigenstates $|+\rangle$ and $|-\rangle$ of the spin observable S_z.

(4.2) *Polar coordinates, Larmor precession*

In the presence of the magnetic field, the coordinates of the unit vector $\vec{u}(t)$ are:

$$\vec{u}(t) = \begin{cases} \theta(t) = \theta = Cte \\ \phi(t) = \omega t + \varphi \end{cases} \qquad [3.108]$$

The angle $\theta(t)$ between the directions of vector $\vec{u}(t)$ and the magnetic field is constant throughout time. On the other hand, $\vec{u}(t)$ and therefore $\vec{M} = M_u \vec{u}$ turns around the field direction with a speed $\omega = d\phi/dt$. This phenomenon is known as Larmor precession.

3.5.4. Solution 4 – Eigenvalue of the square angular momentum operator

l_+ and l_- are defined by the expressions:

$$l_+ = l_x + il_y \; ; \; l_- = l_x - il_y \qquad [3.109]$$

(1) *Proof*

The square angular momentum operator $\vec{l}^{\,2}$ is given by the relation:

$$\vec{l}^{\,2} = l_x^2 + l_y^2 + l_z^2 \qquad [3.110]$$

The product $l_z \vec{l}^{\,2}$ is then written as:

$$l_z \vec{l}^{\,2} = l_z (l_x^2 + l_y^2 + l_z^2)$$

Hence:

$$l_z \vec{l}^{\,2} = (l_z l_x) l_x + (l_z l_y) l_y + l_z^3) \qquad [3.111]$$

Similarly, we obtain:

$$\vec{l}^{\,2} l_z = l_x (l_x l_z) + (l_y (l_y l_z) + l_z^3 \qquad [3.112]$$

Deducing the products $l_x l_z$, $l_y l_z$, $l_z l_x$ and $l_z l_y$ from the commutation relations [2.152] and inserting the obtained results in [3.111] and [3.112], we respectively find:

$$l_z \vec{l}^{\,2} = l_x (l_z l_x - i\hbar l_y) + l_y (l_z l_y + i\hbar l_x) + l_z^3)$$

$$\vec{l}^{\,2} l_z = (l_x l_z + i\hbar l_y) l_x + (l_y l_z - i\hbar l_x) l_y + l_z^3)$$

Subtracting member by member the two equalities above, we have:

$$\vec{l}^{\,2} l_z - l_z \vec{l}^{\,2} = 0 \qquad [3.113]$$

The square angular momentum operator commutes with l_z (it goes without saying that the same commutation relation is valid for the other components l_x and l_y).

CONCLUSION.– Since $\vec{l}^{\,2}$ and l_z commute, there is a physical state in which these two operators have determined values. In other words, these two operators are simultaneously measurable observables.

(2) *Commutators*

Let us calculate the respective commutators $[l_+, l_-]$, $[l_z, l_+]$ and $[l_z, l_-]$.

(2.1) $[l_+, l_-]$

By definition:

$$[l_+, l_-] = l_+l_- - l_-l_+ \qquad [3.114]$$

Using [3.109], we have:

$$l_+l_- = (l_x + il_y)(l_x - il_y)$$
$$l_-l_+ = (l_x - il_y)(l_x + il_y)$$

Hence:

$$l_+l_- = l_x^2 - il_xl_y + il_yl_x + l_y^2$$
$$l_-l_+ = l_x^2 + il_xl_y - il_yl_x + l_y^2$$

Subtracting these equalities member by member, after arrangement we get:

$$l_+ l_- - l_+ l_- = -2i\,(l_xl_y - l_yl_x)$$

Using the commutation relations [2.152], we finally find:

$$l_+l_- - l_-l_+ = 2\hbar l_z \qquad [3.115]$$

(2.2) $[l_z, l_+]$ and $[l_z, l_-]$

Similarly:

$$[l_z, l_+] = l_zl_+ - l_+l_z \qquad [3.116]$$

Using [3.109], we get:

$$l_zl_+ = l_z(l_x + il_y) = l_zl_x + il_zl_y$$
$$l_+l_z = (l_x + il_y)l_z = l_xl_z + il_yl_z$$

Subtracting these equalities member by member, we have:

$$l_zl_+ - l_+l_z = (l_zl_x - l_xl_z) + i(l_zl_y - l_yl_z) \qquad [3.117]$$

Using [2.152], equality [3.117] gives:

$$l_z l_+ - l_+ l_z = i\hbar l_y + \hbar l_x = \hbar(l_x + il_y)$$

Or according to [3.109]:

$$l_z l_+ - l_+ l_z = \hbar l_+ \quad [3.118]$$

Similarly, we find:

$$l_z l_- - l_- l_z = -\hbar l_- \quad [3.119]$$

(3) *Expression, proof*

Let us deduce from [3.109] the expressions of l_x and l_y as a function of l_+ and l_-. Inserting the results obtained in [3.110], we find:

$$\vec{l}^2 = \left(\frac{l_+ + l_-}{2}\right)^2 + \left(\frac{l_+ - l_-}{2i}\right)^2 + l_z^2 \quad [3.120]$$

Expanding this expression, we have:

$$\vec{l}^2 = \frac{1}{4}\left(l_+^2 + l_-^2 + l_+ l_- + l_- l_+\right) - \frac{1}{4}\left(l_+^2 + l_-^2 - l_+ l_- - l_- l_+\right) + l_z^2$$

Hence:

$$\vec{l}^2 = \frac{1}{2}(l_+ l_- + l_- l_+) + l_z^2 \quad [3.121]$$

Using [3.115], we finally find:

$$\vec{l}^2 = l_+ l_- + l_z^2 - \hbar l_z = l_- l_+ + l_z^2 + \hbar l_z \quad [3.122]$$

(4) *Equations with eigenvalues*

The highest value of the magnetic quantum number m_ℓ is ℓ. In the eigenstate $|\Phi\rangle$ common to $\vec{l}^2$ and l_z, the eigenvalue of l_z is therefore $\ell\hbar$. If λ^2 designates the eigenvalue of $\vec{l}^2$, the equations with eigenvalues of operators $\vec{l}^2$ and l_z are written as:

$$\vec{l}^{\,2}|\Phi\rangle = \lambda^2|\Phi\rangle \qquad [3.123]$$
$$l_z|\Phi\rangle = m_{\ell\,\max}\hbar|\Phi\rangle = \ell\hbar|\Phi\rangle$$

(5) *Proof*

Let us put that $l_+|\Phi\rangle$ and $l_-|\Phi\rangle$ are eigenstates of l_z associated, respectively, with the eigenvalues α_+ and α_- to be specified. We have:

$$\begin{cases} l_z l_+|\Phi\rangle = \alpha_+|\Phi\rangle \\ l_z l_-|\Phi\rangle = \alpha_-|\Phi\rangle \end{cases} \qquad [3.124]$$

Using [3.118] and [3.119], we obtain:

$$l_z l_\pm|\Phi\rangle = (l_\pm l_z \pm \hbar l_\pm)|\Phi\rangle$$

Hence:

$$l_z l_\pm|\Phi\rangle = l_\pm l_z|\Phi\rangle \pm \hbar l_\pm|\Phi\rangle$$

Or according to [3.123]:

$$l_z l_\pm|\Phi\rangle = \hbar(\ell \pm 1) l_\pm|\Phi\rangle \qquad [3.125]$$

A comparison between equations [3.124] and [3.125] reveals that vectors $l_+|\Phi\rangle$ and $l_-|\Phi\rangle$ are eigenstates of l_z associated, respectively, with the eigenvalues α_+ and α_- given by:

$$\alpha_+ = \hbar(\ell+1)\,;\, \alpha_- = \hbar(\ell-1) \qquad [3.126]$$

(6) *Incompatible eigenvalue*

The eigenvalue α_+ is incompatible with the principle of quantization of l_z. Indeed, the eigenvalues of l_z are equal to $m_\ell\hbar$. Knowing that the maximal m_ℓ is ℓ, it follows that $\alpha_+ = \hbar(\ell+1)$ cannot be an eigenvalue of l_z.

(7) *Eigenvalue of the square angular momentum operator*

The equation with eigenvalues [3.125] results from the commutation relations [3.117]. Consequently, equation [3.125] is well defined, though it leads to an eigenvalue α_+ that is unacceptable for l_z. To eliminate the contradiction, the condition $l_+|\Phi\rangle = 0$ must be imposed. This condition implies:

$$l_- l_+|\Phi\rangle = 0 \quad [3.127]$$

Using [3.122], we have:

$$\vec{l}^{\,2}|\Phi\rangle = (l_+l_- + l_z^2 + \hbar l_z)|\Phi\rangle$$

Taking equations [3.123] and [3.127] into account, we get:

$$\vec{l}^{\,2}|\Phi\rangle = \hbar^2\ell(\ell+1)|\Phi\rangle \quad [3.128]$$

Comparing the first equation [3.123] and [3.128], it can be noted that the eigenvalue λ^2 of the square angular momentum operator $\vec{l}^{\,2}$ is equal to:

$$\lambda^2 = \hbar^2\ell(\ell+1) \quad [3.129]$$

(8) *Representation*

In the context of the vector diagram model, the angular momentum operator $\vec{l}$ is represented by a vector of module:

$$|\vec{l}| = \sqrt{\vec{l}^{\,2}} = \hbar\sqrt{\ell(\ell+1)} \quad [3.130]$$

For a given state, m_ℓ takes values from $-\ell$ to $+\ell$, hence $(2\ell+1)$ values.

In $\hbar$ unit, the possible values of m_ℓ are figured on the axis *Oz*, being considered projections on this axis of vector $\vec{l}$ of length $\sqrt{\ell(\ell+1)}$.

For $\ell = 2$, the module of vector $\vec{l}$ is $\sqrt{6}$. These results, which determine the various possible values of operators $\vec{l}^{\,2}$ and l_z, are known as *spatial quantization*, which is generally represented in the form of a *vector diagram*, as illustrated in Figure 3.6 for $\ell = 2$.

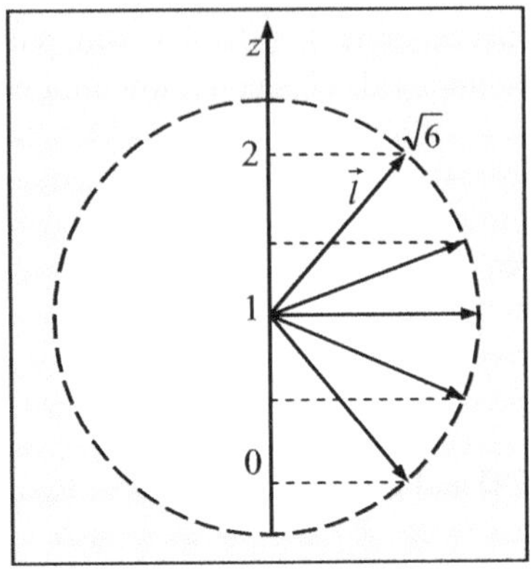

Figure 3.6. *Vector diagram reflecting the spatial quantization of the angular momentum operator*

3.5.5. *Solution 5 – Constant of motion, good quantum numbers*

(1) *Equations with eigenvalues*

The system $\{|\Phi_{n,k,m}\rangle\}$ of eigenkets is common to observables A and H with eigenvalues E_n and a_k, respectively. The discrete index m identifies the eigenvalues of the observables that potentially form a CSCO with H and A. The equations with eigenvalues of observables H and A are then written as follows:

$$\begin{cases} H\left|\Phi_{n,k,m}\right\rangle = E_n\left|\Phi_{n,k,m}\right\rangle \\ A\left|\Phi_{n,k,m}\right\rangle = a_k\left|\Phi_{n,k,m}\right\rangle \end{cases} \quad [3.131]$$

(2) *Constant of motion*

Operator A is explicitly time independent and commutes with Hamiltonian H, which is itself explicitly time independent: therefore, H and A are constants of motion.

(3) *Good quantum numbers*

The states $|\Phi_{n,k,m}\rangle$ of H being stationary states (H is time independent), if the system is in state $|\Phi_{n,k,m}\rangle$, it will remain so indefinitely. Or, according to [3.131], the states $|\Phi_{n,k,m}\rangle$ are also eigenstates of A with eigenvalues a_k. Hence, if A is a constant of motion, there are eigenstates of the system that remain at any instant eigenstates of A with eigenvalues a_k. For this reason, the eigenvalues a_k are known as *good quantum numbers*.

(4) *Expressions of the state vectors*

The expansions of the state vectors $|\Psi(t_0)\rangle$ and $|\Psi(t)\rangle$ on the eigenstates $|\Phi_{n,k,m}\rangle$ are written as, respectively:

$$\left|\Psi(t_0)\right\rangle = \sum_{n,k,m} c_{n,k,m}(t_0)\left|\Phi_{n,k,m}\right\rangle \qquad [3.132]$$

$$\left|\Psi(t)\right\rangle = \sum_{n,k,m} c_{n,k,m}(t)\left|\Phi_{n,k,m}\right\rangle \qquad [3.133]$$

In these expressions, the components $c_{n,k,m}(t)$ are deduced from [3.48]. Hence:

$$c_{n,k,m}(t) = c_{n,k,m}(t_0)\, e^{-iE_n(t-t_0)/\hbar} \qquad [3.134]$$

3.5.6. *Solution 6 – Evolution of the mean values of the operators associated with position and linear momentum*

(1) *Evolution equations, connection with classical mechanics*

According to Ehrenfest's theorem, the equations of evolution of the mean values $\langle X\rangle$ and $\langle P\rangle$ are written as follows:

$$\begin{cases} \dfrac{d\langle X\rangle}{dt} = \dfrac{\langle P\rangle}{m} \\ \dfrac{d\langle P\rangle}{dt} = -\langle \nabla V(X)\rangle \end{cases} \qquad [3.135]$$

For a free particle, $V(X) = 0$. Therefore, according to [3.135]:

$$\begin{cases} \dfrac{d}{dt}\langle X\rangle = \dfrac{1}{m}\langle P\rangle \\ \dfrac{d}{dt}\langle P\rangle = 0 \end{cases} \qquad [3.136]$$

System [3.136] leads to the following result:

$$\langle P\rangle = \langle P\rangle_0 = \text{Cst} \qquad [3.137]$$

Hence, the first of equations [3.136] is written as:

$$\frac{d}{dt}\langle X\rangle=\frac{1}{m}\langle P\rangle_0$$

Therefore:

$$\langle X\rangle=\frac{\langle P\rangle_0}{m}t+\langle X\rangle_0 \qquad [3.138]$$

The classical equivalent of equation [3.138] is the equation of motion of a vehicle of mass m in uniform rectilinear motion: $x\ (t)\ =\ v_0\ +\ x_0$. Knowing that $p = mv$, this leads to:

$$x(t)=\frac{p}{m}t+x_0 \qquad [3.139]$$

The classical equation [3.139] is quite similar to the quantum equation [3.138].

(2) *Evolution equations, connection with classical mechanics*

The observables X and P are time independent.

– *For* $\langle P^2\rangle$

$$\frac{d}{dt}\left\langle P^2\right\rangle=\frac{1}{i\hbar}\left\langle[P^2,H]\right\rangle \qquad [3.140]$$

The commutator:

$$[P^2,H]=[P^2,\frac{P^2}{2m}+V(X)]=[P^2,V(X)]$$

Or:

$$\begin{aligned}[P^2,V(X)]&=P[P,V(X)]+[P,V(X)]P\\&=-2i\hbar P\frac{\partial V(X)}{\partial X}\end{aligned}$$

For a free particle, $V(X) = 0$. Hence:

$$[P^2,H]=[P^2,V(X)]=0 \qquad [3.141]$$

Equation [3.140] is then written as:

$$\frac{d}{dt}\langle P^2 \rangle = 0 \Rightarrow \langle P^2 \rangle = \langle P^2 \rangle_0 = \text{Cst} \qquad [3.142]$$

– *For* $\langle X^2 \rangle$

$$\frac{d}{dt}\langle X^2 \rangle = \frac{1}{i\hbar}\langle [X^2, H] \rangle \qquad [3.143]$$

Let us determine the commutator in [3.143]. Knowing that $[X^2, V(X)] = 0$, we obtain:

$$[X^2, H] = [X^2, \frac{P^2}{2m} + V(X)] = [XX, \frac{P^2}{2m}] \qquad [3.144]$$

Or:

$$[XX, \frac{P^2}{2m}] = \frac{1}{2m}\left(X[X, P^2] + [X, P^2]X\right)$$

Knowing that $[X, P^n] = i\hbar n P^{n-1}$ (see result [2.62]), the last commutator in [3.144] is written as:

$$[XX, \frac{P^2}{2m}] = \frac{i\hbar}{m}(XP + PX) \qquad [3.145]$$

Using result [3.145], equation [3.143] can then be written as:

$$\frac{d}{dt}\langle X^2 \rangle = \frac{1}{m}\langle XP + PX \rangle \qquad [3.146]$$

The integration of equation [3.146] requires the calculation of $\langle XP + PX \rangle$.

Considering the equation of evolution of the mean $\langle XP + PX \rangle$, according to Ehrenfest's theorem we obtain:

$$\frac{d}{dt}\langle XP + PX \rangle = \frac{1}{i\hbar}[XP + PX, H] \qquad [3.147]$$

The commutator in the right member of [3.147] is:

$$[XP + PX, H] = [XP, H] + [PX, H]$$

which is:

$$[XP + PX, H] = X[P, H] + [X, H]P + P[X, H] + [P, H]X \qquad [3.148]$$

For a free particle, $V(X) = 0$ and $H = P^2/2m$. Therefore, [3.148] becomes:

$$[XP + PX, H] = \frac{1}{2m}X[P,P^2] + \frac{1}{2m}[X,P^2]P + \\ + \frac{1}{2m}P[X,P^2] + \frac{1}{2m}[P,P^2]X$$

which is:

$$[XP + PX, H] = \frac{1}{m}i\hbar P^2 + \frac{1}{m}i\hbar P^2$$

or:

$$[XP + PX, H] = \frac{2i\hbar}{m}P^2 \qquad [3.149]$$

Using result [3.149], equation [3.147] is written as:

$$\frac{d}{dt}\langle XP + PX\rangle = \frac{2}{m}\langle P^2\rangle \qquad [3.150]$$

Taking [3.142] into account, equation [3.150] is written as:

$$\frac{d}{dt}\langle XP + PX\rangle = \frac{2}{m}\langle P^2\rangle_0 \qquad [3.151]$$

The integration of this equation leads to:

$$\langle XP + PX\rangle = \frac{2}{m}\langle P^2\rangle_0 t + \langle XP + PX\rangle_0 \qquad [3.152]$$

Using [3.152], equation [3.146] can be written in the form:

$$\frac{d}{dt}\left\langle X^2\right\rangle = \frac{2}{m^2}\left\langle P^2\right\rangle_0 t + \frac{1}{m}\langle XP + PX\rangle_0$$

Or after integration:

$$\left\langle X^2\right\rangle(t) = \frac{1}{m^2}\left\langle P^2\right\rangle_0 t^2 + \frac{1}{m}\langle XP + PX\rangle_0 t + \left\langle X^2\right\rangle_0 \qquad [3.153]$$

The mean value [3.153] evolves according to the classical law of uniformly accelerated motion along the axis *Ox*:

$$x(t) = \frac{1}{2}a\,t^2 + v_0 t + x_0$$

(3) *Root mean square deviation*

By definition, the root mean square deviation ΔX is defined by the following relation:

$$\Delta X = \sqrt{\left\langle X^2\right\rangle - \langle X\rangle^2} \qquad [3.154]$$

Using [3.138] and [3.153], we obtain according to [3.154]:

$$(\Delta X)^2 = \frac{1}{m^2}\left\langle P^2\right\rangle_0 t^2 + \frac{1}{m}\langle XP + PX\rangle_0 t + \left\langle X^2\right\rangle_0 - \left(\frac{1}{m}\langle P\rangle_0 t + \langle X\rangle_0\right)^2 \qquad [3.155]$$

Knowing that at $t = 0$, $\langle XP + PX\rangle_0 = \langle X\rangle_0$, equation [3.155] becomes:

$$(\Delta X)^2 = \frac{1}{m^2}\left\langle P^2\right\rangle_0 t^2 + \left\langle X^2\right\rangle_0 - \frac{1}{m^2}\langle P\rangle_0^2 t^2 \qquad [3.156]$$

Let us insert into the first term of the right member of equation [3.156] the mean $\langle X\rangle_0^2$, though it is zero. We obtain:

$$(\Delta X)^2 = \frac{1}{m^2}\left(\left\langle P^2\right\rangle_0 - \langle P\rangle_0^2\right)t^2 + \left(\left\langle X^2\right\rangle_0 - \langle X\rangle_0^2\right)$$

It can be noted that the second member of this relation contains the square of root mean square deviations $(\Delta X)_{t=0}$ and $(\Delta P)_{t=0}$. Hence:

$$(\Delta X)^2 = \frac{1}{m^2}(\Delta P)_0^2\, t^2 + (\Delta X)_0^2$$

or:

$$\Delta X = \frac{1}{m}\sqrt{(\Delta P)_0^2\, t^2 + (\Delta X)_0^2} \qquad [3.157]$$

CONCLUSION.– When t increases, the width ΔX of the wave packet increases: this reflects the packet spreading.

3.5.7. *Solution 7 – Particle subjected to various potentials*

(1) *Evolution equations*

According to Ehrenfest's theorem, the mean values $\langle X\rangle$ and $\langle P\rangle$ evolve according to the laws [3.135]. The potential to which a particle is submitted has the following form: $V(x) = ax^n$.

(1.1) *Free particle*

For a free particle, the potential is zero. We obtain [3.137] and [3.138].

(1.2) *Particle subjected to uniform potential*

For a particle subjected to a uniform potential: $V(x) = a$ $(n = 0)$ = Cst. Equations [3.137] and [3.138] are still verified since $\nabla V(X) = 0$.

(1.3) *Particle subjected to parabolic potential*

For a particle subjected to a parabolic potential, $V(x) = ax^2$. Using [3.135], we find knowing that $\nabla V(X) = 2aX$:

$$\begin{cases} \dfrac{d}{dt}\langle X\rangle = \dfrac{1}{m}\langle P\rangle \\ \dfrac{d}{dt}\langle P\rangle = -2a\langle X\rangle \end{cases} \qquad [3.158]$$

Using equations [3.158], we have:

$$\frac{d^2}{dt^2}\langle X\rangle = \frac{1}{m}\frac{d}{dt}\langle P\rangle \qquad [3.159]$$

Taking the second of equations [3.158] into account, equation [3.159] can be written as:

$$\frac{d^2}{dt^2}\langle X\rangle + \frac{2a}{m}\langle X\rangle = 0 \qquad [3.160]$$

The integration of this equation finally leads to:

$$\langle X\rangle(t) = \langle X\rangle_0 \cos(\omega_0 t + \varphi_0) \qquad [3.161]$$

where $\omega_0 = \sqrt{2a/m}$.

(2) *Connection with classical mechanics*

The differential equation of a classical harmonic oscillator can be written as:

$$m\frac{d^2x}{dt^2} = -kx \qquad [3.162]$$

The classical equation [3.162] is the equivalent of the quantum equation [3.160]. Moreover, the solution to the differential equation [3.162] has the well-known form:

$$x(t) = x_0 \cos(\omega_0 t + \varphi_0)$$

$$\text{or } \omega_0 = \sqrt{k/m} \qquad [3.163]$$

Comparing the laws [3.163] and [3.161], it can be seen that the center of the wave packet behaves as a classical particle.

(3) *Behavior of the particle*

For $V(x) = ax^3$, using [3.135] and knowing that $\nabla V(X) = 3aX$, we obtain:

$$\begin{cases} \dfrac{d\langle X\rangle}{dt} = \dfrac{\langle P\rangle}{m} \\ \dfrac{d\langle P\rangle}{dt} = -3a\langle X^2\rangle \end{cases}$$

which then yields:

$$\frac{d^2}{dt^2}\langle X\rangle = -\frac{3a}{m}\langle X^2\rangle \qquad [3.164]$$

Equation [3.164] has no classical equivalent. Consequently, the center of the wave packet does not behave as a classical particle.

GENERAL CONCLUSION.– The center of the wave packet does not rigorously follow the laws of classical mechanics for the free particle and for the particle subjected to uniform or parabolic potential.

3.5.8. *Solution 8 – Oscillating molecular dipole, root mean square deviation*

The potential energy of the HCl dipole has the form:

$$V(x) = \frac{1}{2}(x-a)^2 \qquad [3.165]$$

The normed wave functions $\Phi_0(x)$ of the ground state and $\Phi_1(x)$ of the first excited state are given by the expressions:

$$\Phi_0(x) = \left(\frac{m\omega}{\pi\hbar}\right)^{\frac{1}{4}} e^{-\frac{1}{2}\frac{m\omega}{\hbar}(x-a)^2} \qquad [3.166]$$

$$\Phi_1(x) = \left[\frac{4}{\pi}\left(\frac{m\omega}{\hbar}\right)^3\right]^{\frac{1}{4}}(x-a)e^{-\beta^2(x-a)^2/2} \qquad [3.167]$$

(1) *Mean value*

In the general case, the kets $|\Phi_n\rangle$ being normed, the mean value $\langle X\rangle$ is:

$$\langle X\rangle_n = \langle\Phi_n|X|\Phi_n\rangle \qquad [3.168]$$

Using the closing relation in $\{|x\rangle\}$ representation, we have:

$$\langle X\rangle_n = \langle\Phi_n|\int_{-\infty}^{+\infty} dx|x\rangle\langle x|X|\Phi_n\rangle \qquad [3.169]$$

Hence:

$$\langle X\rangle_n = \int_{-\infty}^{+\infty} dx\langle\Phi_n|x\rangle x\langle x|\Phi_n\rangle$$

Therefore:

$$\langle X\rangle_n = \int_{-\infty}^{+\infty} dx\, x|\Phi_n(x)|^2 \qquad [3.170]$$

– *Case of the ground state*

Considering [3.166], relation [3.170] gives:

$$\langle X\rangle_0 = \left(\frac{m\omega}{\pi\hbar}\right)^{\frac{1}{2}} \int_{-\infty}^{+\infty} dx\ x\ e^{-\frac{m\omega}{\hbar}(x-a)^2} \qquad [3.171]$$

Putting:

$$y = x - a, \text{ and } A = \left(\frac{m\omega}{\pi\hbar}\right)^{\frac{1}{2}},\ \rho = \frac{m\omega}{\hbar} \qquad [3.172]$$

then equation [3.171] is written in the form:

$$\langle X\rangle_0 = A\left\{\int_{-\infty}^{+\infty} dy\ y\ e^{-\rho y^2} + 2a\int_{0}^{+\infty} dy\ e^{-\rho y^2}\right\} \qquad [3.173]$$

Knowing that $ye^{-\rho y^2}$ is an odd function, then:

$$\int_{-\infty}^{+\infty} dy\ y e^{-\rho y^2} = 0$$

Using [3.172], equation [3.173] yields:

$$\langle X \rangle_0 = 2AaI_0 = Aa\sqrt{\frac{\pi}{\rho}} = a \times \left(\frac{m\omega}{\pi\hbar}\right)^{\frac{1}{2}} \times \left(\frac{\pi\hbar}{m\omega}\right)^{\frac{1}{2}}$$

or finally:

$$\langle X \rangle_0 = a \qquad [3.174]$$

– *Case of the first excited state*

Considering the wave function [3.167], relation [3.170] yields:

$$\langle X \rangle_1 = \left[\frac{4}{\pi}\left(\frac{m\omega}{\hbar}\right)^3\right]^{\frac{1}{2}} \int_{-\infty}^{+\infty} dx\, x(x-a)^2 \times e^{-\frac{m\omega}{\hbar}(x-a)^2} \qquad [3.175]$$

Let us use the changes of variables [3.172] where:

$$A = \left[\frac{4}{\pi}\left(\frac{m\omega}{\hbar}\right)^3\right]^{\frac{1}{2}}$$

Equation [3.175] then gives (with $y = x - a$):

$$\langle X \rangle_1 = A\left\{\int_{-\infty}^{+\infty} dy\, y^3 e^{-\rho y^2} + a\int_{-\infty}^{+\infty} dy\, y^2 e^{-\rho y^2}\right\}$$

That means:

$$\langle X \rangle_1 = A\left\{\int_{-\infty}^{+\infty} dy\, y^3 e^{-\rho y^2} + 2a\int_{-\infty}^{+\infty} dy\, y^2 e^{-\rho y^2}\right\} \qquad [3.176]$$

Knowing that $y^3 e^{-\rho y^2}$ is an odd function, we have:

$$\int_{-\infty}^{+\infty} dy\, y^3 e^{-\rho y^2} = 0$$

Then, using [3.176] we obtain:

$$\langle X \rangle_1 = 2AaI_2 = Aa\frac{1}{\rho}I_0 = a\,A\sqrt{\frac{\pi}{4\rho^3}}$$

Replacing A and ρ by their expressions, we get:

$$\langle X \rangle_1 = a \times \left[\frac{4}{\pi}\left(\frac{m\omega}{\hbar}\right)^3\right]^{\frac{1}{2}} \times \left[\frac{\pi}{4}\left(\frac{\hbar}{m\omega}\right)^3\right]^{\frac{1}{2}}$$

or finally:

$$\langle X \rangle_1 = a. \qquad [3.177]$$

NOTE.– Results [3.174] and 3.177] are identical. This is justified in the context of the adopted model, which does not reflect reality, as the potential is not perfectly harmonic.

(2) *Uncertainty in the measurement of the position of the hydrogen atom*

$$\Delta X = \sqrt{\langle \Phi | (X-a)^2 | \Phi \rangle} \qquad [3.178]$$

(2.1) *Calculation of uncertainties*

Using the closing relation satisfied in $\{|x\rangle\}$ representation, the square of the uncertainty $(\Delta X)_n$ is written according to [3.178]:

$$(\Delta X)_n^2 = \int_{-\infty}^{+\infty} dx\,(x-a)^2 |\Phi_n(x)|^2 \qquad [3.179]$$

– *Case of the ground state*

Using [3.166], we get, according to [3.179]:

$$(\Delta X)_0^2 = \left(\frac{m\omega}{\pi\pi}\right)^{\frac{1}{2}} \int_{-\infty}^{+\infty} dx\,(x-a)^2 e^{\frac{-m\omega}{\hbar}(x-a)^2} \quad [3.180]$$

Using the same changes of variables [3.180], we have:

$$(\Delta X)_0^2 = 2A \int_0^{+\infty} dy\, y^2 e^{-\rho y^2} = 2AI_2 = \frac{A}{\rho} I_0$$

Hence:

$$(\Delta X)_0^2 = A\sqrt{\frac{\pi}{4\rho^3}} = \left(\frac{m\omega}{\pi\hbar}\right)^{\frac{1}{2}} \times \left(\frac{\hbar}{m\omega}\right)^{\frac{3}{2}} \times \left(\frac{\pi}{4}\right)^{\frac{1}{2}}$$

or:

$$(\Delta X)_0^2 = \frac{1}{2}\frac{\hbar}{m\omega}$$

which finally leads to:

$$(\Delta X)_0 = \sqrt{\frac{1}{2}\frac{\hbar}{m\omega}} \quad [3.181]$$

– *Case of the first excited state*

Using [3.167], equation [3.179] yields:

$$(\Delta X)_1^2 = A \int_{-\infty}^{+\infty} dx\,(x-a)^4 e^{-\frac{m\omega}{\hbar}(x-a)^2} \quad [3.182]$$

Using the same changes of variables as previously, we have:

$$(\Delta X)_1^2 = 2A \int_0^{+\infty} dy\, y^4 e^{-\rho y^2} = 2AI_4 = \frac{3A}{\rho} I_2$$

which is:

$$(\Delta X)_1^2 = \frac{3A}{2\rho^2} I_0 = \frac{3A}{4}\sqrt{\frac{\pi}{\rho^5}}$$

Replacing A and ρ by their expressions [6.328], we have:

$$(\Delta X)_1^2 = \frac{3}{4} \times \left[\frac{4}{\pi}\left(\frac{m\omega}{\hbar}\right)^3\right]^{\frac{1}{2}} \times \left(\frac{\hbar}{m\omega}\right)^{\frac{5}{2}} \times \sqrt{\pi}$$

Hence:

$$(\Delta X)_1^2 = \frac{3}{2} \times \left(\frac{\hbar}{m\omega}\right)$$

or finally:

$$(\Delta X)_1 = \sqrt{\frac{3}{2}\frac{\hbar}{m\omega}} \qquad [3.183]$$

NOTE.– Integrating the general equation [3.179], we find:

$$(\Delta X)_n^2 = \left(n + \frac{1}{2}\right)\frac{\hbar}{m\omega} \qquad [3.184]$$

Result [3.184] actually gives the specific cases [3.181] and [3.183], respectively, for the ground state ($n = 0$) and for the first excited state ($n = 1$).

(2.2) *Interpretation*

Results [3.181] and [3.183] show that the uncertainty $(\Delta X)_n$ increases with the excitation (when n increases). This becomes clearer when the general expression [3.184] is considered.

3.5.9. *Solution 9 – Infinite potential well, time–energy uncertainty relation*

The particle is confined in an infinite potential well of width l, such that:

$V(x) = 0, \text{ if } 0 \leq x \leq l$

$V(x) = \infty, \text{ elsewhere}$

The Hamiltonian of the particle is:

$$H = \frac{P^2}{2m} + V(X) \qquad [3.185]$$

The eigenvalues of the Hamiltonian H are given by the expression:

$$E_n = \frac{\hbar^2 \pi^2}{2ml^2} n^2 \qquad [3.186]$$

The eigenfunctions of H are such that:

$$\Phi_n(x) = \sqrt{\frac{2}{l}} \sin\left(\frac{n\pi x}{l}\right), \text{ if } 0 < x < l$$

$$\Phi_n(x) = 0, \text{ elsewhere.} \qquad [3.187]$$

At instant $t = 0$, the state of the particle is described by the ket:

$$|\Psi(0)\rangle = \frac{1}{\sqrt{2}}(|\Phi_1\rangle + |\Phi_2\rangle) \qquad [3.188]$$

(1) *Expression of the wave function*

Using [3.188], we get in $\{|x\rangle\}$ representation:

$$\langle x|\Psi(0)\rangle = \frac{1}{\sqrt{2}}(\langle x|\Phi_1\rangle + \langle x|\Phi_2\rangle)$$

Hence:

$$\Psi(x) = \frac{1}{\sqrt{2}}[\Phi_1(x) + \Phi_2(x)] \qquad [3.189]$$

Using [3.187], the wave function [3.189] is written as:

$$\Psi(x) = \frac{1}{\sqrt{l}}\left\{\sin\left(\frac{\pi x}{l}\right) + \sin\left(\frac{2\pi x}{l}\right)\right\} \quad [3.190]$$

(2) *Proof*

At instant t, the state vector is given by the expression:

$$\left|\Psi(t)\right\rangle = \sum_{n=1}^{2} c_n(0)e^{-iE_n t/\hbar}\left|\Phi_n\right\rangle \quad [3.191]$$

Hence:

$$\left|\Psi(t)\right\rangle = c_1(0)e^{-iE_1 t/\hbar}\left|\Phi_1\right\rangle + c_2(0)e^{-iE_2 t/\hbar}\left|\Phi_2\right\rangle \quad [3.192]$$

This expression is valid for any t. In particular, at $t = 0$, we have:

$$\left|\Psi(0)\right\rangle = c_1(0)\left|\Phi_1\right\rangle + c_2(0)\left|\Phi_2\right\rangle \quad [3.193]$$

Comparing [3.189] and [3.193], we see that:

$$c_1(0) = c_2(0) = \frac{1}{\sqrt{2}} \quad [3.194]$$

Moreover, according to [3.186]:

$$E_1 = \frac{\hbar^2\pi^2}{2ml^2}\ ;\ E_2 = 2\frac{\hbar^2\pi^2}{2ml^2} \quad [3.195]$$

Using [3.194] and [3.195], expression [3.192] is written in the following form:

$$\left|\Psi(t)\right\rangle = \frac{1}{\sqrt{2}}\left\{e^{-i\frac{\hbar\pi^2}{2ml^2}t}\left|\Phi_1\right\rangle + e^{-i\frac{\hbar\pi^2}{ml^2}t}\left|\Phi_2\right\rangle\right\} \quad [3.196]$$

We put:

$$\alpha_1 = \frac{\hbar\pi^2}{2ml^2}\ ;\ \alpha_2 = \frac{\hbar\pi^2}{ml^2} \quad [3.197]$$

Taking [3.198] into account, the wave function [3.196] is written in the form:

$$|\Psi(t)\rangle = \frac{1}{\sqrt{2}}\left[e^{-i\alpha_1 t}|\Phi_1\rangle + e^{-i\alpha_2 t}|\Phi_2\rangle\right] \quad [3.198]$$

(3) *Density of probability of presence*

The density of the probability of presence $D(x, t)$ of the particle is written as:

$$D(x,t) = |\Psi(x,t)|^2 \quad [3.199]$$

Considering [6.354], we get in $\{|x\rangle\}$ representation:

$$\Psi(x,t) = \frac{1}{\sqrt{2}}\left[e^{-iE_1 t/\hbar}\Phi_1(x) + e^{-iE_2 t/\hbar}\Phi_2(x)\right] \quad [3.200]$$

The complex conjugate of the wave function [3.200] is written as:

$$\Psi^*(x,t) = \frac{1}{\sqrt{2}}\left[e^{iE_1 t/\hbar}\Phi_1(x) + e^{iE_2 t/\hbar}\Phi_2(x)\right] \quad [3.201]$$

Using expressions [3.200] and [3.201], the density of probability is written as:

$$D(x,t) = \frac{1}{2}\Phi_1^2(x) + \frac{1}{2}\Phi_2^2(x) + \Phi_1(x)\Phi_2(x)\left[e^{-iE_1 t/\hbar}e^{iE_2 t/\hbar} + e^{iE_1 t/\hbar}e^{-iE_2 t/\hbar}\right]$$

Hence:

$$D(x,t) = \frac{1}{2}\Phi_1^2(x) + \frac{1}{2}\Phi_2^2(x) + \Phi_1(x)\Phi_2(x)\left[e^{i(E_2-E_1)t/\hbar} + e^{-i(E_2-E_1)t/\hbar}\right] \quad [3.202]$$

We put:

$$\omega_{21} = \frac{(E_2 - E_1)}{\hbar} \quad [3.203]$$

Taking [3.203] into account, expression [3.202] is written in the following form:

$$D(x,t) = \frac{1}{2}\Phi_1^2(x) + \frac{1}{2}\Phi_2^2(x) + \Phi_1(x)\Phi_2(x)\left[e^{i\omega_{21}t} + e^{-i\omega_{21}t}\right]$$

Hence:

$$D(x,t) = \frac{1}{2}\left\{\Phi_1^2(x) + \frac{1}{2}\Phi_2^2(x) + 2\Phi_1(x)\Phi_2(x)\cos(\omega_{21}t)\right\} \quad [3.204]$$

CONCLUSION.– Expression [3.204] shows that the variation in time of the density of probability is due to the interference term, which is proportional to the product $\Phi_1(x)\,\Phi_2(x)$. This term, which is responsible for the evolution of the wave packet, oscillates in time with the single Bohr angular frequency given by [3.203].

NOTE.– Figure 3.7 indicates the plots of the density of the probability of presence for the ground state, for the first excited state and for the interference term responsible for the evolution of the shape of the wave packet.

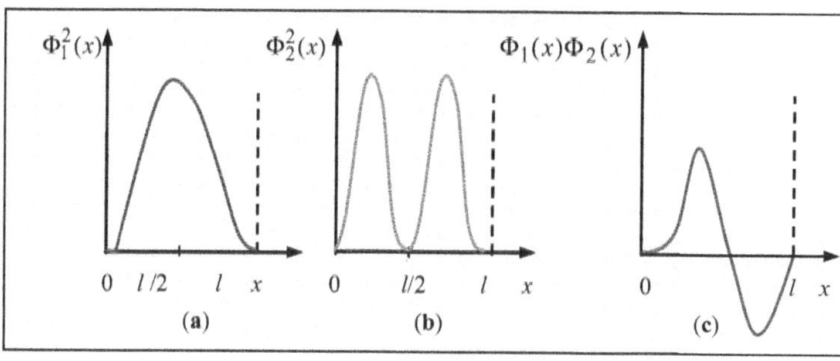

Figure 3.7. *Plots of the densities of probability of presence of the particle $\Phi_1^2(x)$ for the ground state (a) and $\Phi_2^2(x)$ for the first excited state (b) and of the interference term Φ_1 (x) Φ_2 (x) (c) responsible for the evolution of the wave packet*

(4) *Mean value*

Taking [3.199] into account, the mean value $\langle X\rangle(t)$ representing the motion of the center of the wave packet associated with the particle is given by the following equation:

$$\langle X\rangle(t) = \langle\Psi(t)|X|\Psi(t)\rangle = \int_{-\infty}^{+\infty} dx\, x|\Psi(x,t)|^2 = \int_{-\infty}^{+\infty} dx\, x\, D(x,t) \quad [3.205]$$

Using [3.204], relation [3.205] yields:

$$\langle X\rangle(t) = \frac{1}{2}\int_{-\infty}^{+\infty} dx\, x\Phi_1^2(x) + \frac{1}{2}\int_{-\infty}^{+\infty} dx\, x\Phi_2^2(x) + \int_{-\infty}^{+\infty} dx\, x\,\Phi_1(x)\Phi_2(x)\cos(\omega_{21}t) \quad [3.206]$$

Knowing that the wave function does not differ from zero unless $0 < x < l$, [3.206] is written as:

$$\langle X\rangle(t) = \frac{1}{2}\int_0^l dx\, x\Phi_1^2(x) + \frac{1}{2}\int_0^l dx\, x\Phi_2^2(x) + \int_0^l dx\, x\Phi_1(x)\Phi_2(x)\cos(\omega_{21}t) \qquad [3.207]$$

To facilitate the integration of [3.207], we put $X' = X - l/2$ for symmetry reasons. This leads to:

$$\langle X\rangle(t) = \langle X'\rangle(t) + \frac{l}{2} \qquad [3.208]$$

The mean value of the observable X' is equal to:

$$\langle X'\rangle(t) = \langle \Psi(t)|X - l/2|\Psi(t)\rangle = \int_{-\infty}^{+\infty} dx\left(x - \frac{l}{2}\right)D(x,t) \qquad [3.209]$$

According to [3.199], this yields:

$$\langle X'\rangle(t) = \frac{1}{2}\int_0^l dx\left(x - \frac{l}{2}\right)\Phi_1^2(x) + \frac{1}{2}\int_0^l dx\left(x - \frac{l}{2}\right)\Phi_2^2(x) + \int_0^l dx\left(x - \frac{l}{2}\right)\Phi_1(x)\Phi_2(x)\cos(\omega_{21}t) \qquad [3.210]$$

Deducing the wave functions $\Phi_1(x)$ and $\Phi_2(x)$ from the general expression [3.187] and inserting the resulting expressions in [3.210], we get:

$$\langle X'\rangle(t) = \frac{1}{l}\int_0^l dx\left(x - \frac{l}{2}\right)\sin^2\left(\frac{\pi x}{l}\right) + \frac{1}{l}\int_0^l dx\left(x - \frac{l}{2}\right)\sin^2\left(\frac{2\pi x}{l}\right) + \frac{2}{l}\int_0^l dx\left(x - \frac{l}{2}\right)\sin\left(\frac{\pi x}{l}\right)\sin\left(\frac{2\pi x}{l}\right)\cos(\omega_{21}t) \qquad [3.211]$$

The two first integrals of [3.211] are zero, as they are of the type:

$$\int_0^a dq\left(q-\frac{a}{2}\right)\sin^2\left(\frac{n\pi q}{a}\right)=0$$

We then obtain:

$$\langle X'\rangle(t)=\frac{2}{l}\cos(\omega_{21}t)\int_0^l dx\left(x-\frac{l}{2}\right)\sin\left(\frac{\pi x}{l}\right)\sin\left(\frac{2\pi x}{l}\right)$$

Hence:

$$\langle X'\rangle(t)=\frac{2}{l}\cos(\omega_{21}t)\left\{\int_0^l dx\,x\sin\left(\frac{\pi x}{l}\right)\sin\left(\frac{2\pi x}{l}\right)-\frac{l}{2}\int_0^l dx\sin\left(\frac{\pi x}{l}\right)\sin\left(\frac{2\pi x}{l}\right)\right\} \quad [3.212]$$

The last term of equation [3.212] is zero. It can be more easily verified without integration, as shown below.

Knowing that $(|\Phi_1\rangle,|\Phi_2\rangle)$ constitutes a basis in the space of states, then the two kets are orthonormal. Hence, using the closing relation satisfied by the continuous set $\{|x\rangle\}$: $\langle\Phi_1|\Phi_2\rangle=0$. We then obtain:

$$\langle\Phi_1|\int_{-\infty}^{+\infty}dx|x\rangle\langle x|\Phi_2\rangle=0$$

Hence:

$$\int_{-\infty}^{+\infty}dx\Phi_1(x)\Phi_2(x)=0$$

or finally:

$$\frac{2}{l}\int_0^l dx\sin\left(\frac{\pi x}{l}\right)\sin\left(\frac{2\pi x}{l}\right)=0$$

Considering this result, expression [3.212] is then written as:

$$\langle X'\rangle(t) = \frac{2}{l}\cos(\omega_{21}t)\int_0^l dx\, x \sin\left(\frac{\pi x}{l}\right)\sin\left(\frac{2\pi x}{l}\right) \quad [3.213]$$

The integral in relation [3.213] is of the type:

$$\int_0^a dq \sin\left(\frac{\pi q}{a}\right)\sin\left(\frac{2\pi q}{a}\right) = -\frac{8}{9}\frac{a^2}{\pi^2}$$

which then yields:

$$\langle X'\rangle(t) = -\frac{16l}{9\pi^2}\cos\omega_{21}t \quad [3.214]$$

Taking [3.208] into account, the mean value $\langle X\rangle(t)$ of the observable X representing the motion of the center of the wave packet associated with the particle is then:

$$\langle X\rangle(t) = \frac{l}{2} - \frac{16l}{9\pi^2}\cos\omega_{21}t \quad [3.215]$$

(5) *Graphical representation*

The variations of the mean value $\langle X\rangle(t)$ compared to the law of motion of a classical particle in a back-and-forth motion in the well with the same angular frequency ω_{21} are indicated in Figure 3.8. For the classical particle, the law of motion is reflected by the equation:

$$x(t) = \begin{cases} +\dfrac{l\omega_{12}}{\pi}\cdot t & ; 0 \le t \le \dfrac{\pi}{\omega_{12}} \\ -\dfrac{l\omega_{12}}{\pi}\cdot t + 2l & ; \dfrac{\pi}{\omega_{12}} \le t \le \dfrac{2\pi}{\omega_{12}} \end{cases} \quad [3.216]$$

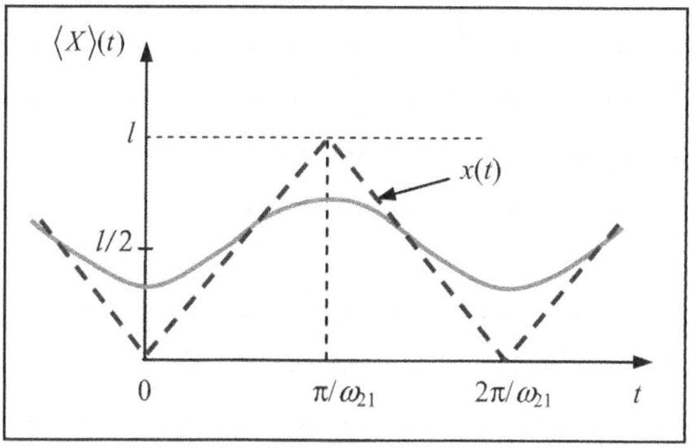

Figure 3.8. *Variation of the mean value ⟨X⟩(t) representing the motion of the center of the wave packet compared to the law of motion of a classical particle that is moving back and forth in the well with the same angular frequency* ω_{21}

Figure 3.8 shows how the classical particle turns back after having reached the higher limit of the well in $x = l$. The quantum mechanics predictions are very different: the center of the wave packet makes a half turn before the particle even gets to $x = l$. This can be explained by the sudden variations of potential at the well boundaries where it passes from zero to infinity: the potential then acts on the packet in such a way that it makes a half-turn before reaching the boundaries $x = 0$ and $x = l$.

(6) *Mean values, root mean square deviation*

– *Mean values*

The mean values $\langle H\rangle(t)$ and $\langle H^2\rangle(t)$ are written as:

$$\langle H\rangle(t) = \langle\psi(t)|H|\psi(t)\rangle = \frac{1}{2}\left[\langle\Phi_1|H|\Phi_1\rangle + \langle\Phi_2|H|\Phi_2\rangle\right] \qquad [3.217]$$

$$\left\langle H^2\right\rangle(t) = \langle\psi(t)|H^2|\psi(t)\rangle = \frac{1}{2}\left[\langle\Phi_1|H^2|\Phi_1\rangle + \langle\Phi_2|H^2|\Phi_2\rangle\right] \qquad [3.218]$$

Knowing that the eigenkets $|\Phi_n\rangle$ are orthonormal, and using the equation with eigenvalues $H|\Phi_n = E_n|\Phi_n\rangle$ we get:

$$\begin{cases} \langle H\rangle(t) = \dfrac{E_1}{2} + \dfrac{E_2}{2} = \dfrac{E_1}{2} + \dfrac{4E_1}{2} = \dfrac{5}{2}E_1 \\ \left\langle H^2\right\rangle(t) = \dfrac{E_1^2}{2} + \dfrac{E_2^2}{2} = \dfrac{E_1^2}{2} + \dfrac{16E_1^2}{2} = \dfrac{17}{2}E_1^2 \end{cases} \qquad [3.219]$$

– *Root mean square deviation ΔH*

Using [3.219], the root mean square deviation ΔH is equal to:

$$\Delta H = \sqrt{\left\langle H^2 \right\rangle(t) - \left\langle H \right\rangle^2(t)} = \sqrt{\frac{17}{2} - \frac{25}{4}} E_1 = \frac{3}{2} E_1 \qquad [3.220]$$

The mean values [3.219] and the root mean square deviation [3.220] are time independent. This is due to the fact that the Hamiltonian of the system is a *constant of motion*.

(7) *Time–energy uncertainty relation*

Taking [3.195] into account, the Bohr angular frequency is:

$$\omega_{21} = \frac{(E_2 - E_1)}{\hbar} = \frac{3E_1}{\hbar} \qquad [3.221]$$

As shown in Figure 3.8, the wave packet evolves significantly after a time period of about $1/\omega_{21}$. Let us put $\Delta t \approx 1/\omega_{21}$. Using [3.221], we get:

$$\Delta t = \approx \frac{\hbar}{3E_1} \qquad [3.222]$$

Using [3.320] and [3.322], we get:

$$\Delta H \Delta t = \frac{3}{2} E_1 \times \frac{\hbar}{3E_1} = \frac{\hbar}{2} \qquad [3.223]$$

Result [3.223] satisfies the lower limit of Heisenberg's time–energy uncertainty relation.

3.5.10. *Solution 10 – Study of a conservative system*

The equation with eigenvalues of the Hamiltonian H_0 is written as follows:

$$H_0 |\Phi_n\rangle = E_0 |\Phi_n\rangle \qquad [3.224]$$

Operator W generating a coupling between various eigenkets $\{|\Phi_n\rangle\}$ of the Hamiltonian H_0 satisfies the following conditions:

$$\langle \Phi_1 | W | \Phi_3 \rangle = \langle \Phi_3 | W | \Phi_1 \rangle = \langle \Phi_1 | W | \Phi_2 \rangle = \langle \Phi_2 | W | \Phi_1 \rangle = \frac{\lambda}{\sqrt{2}} \qquad [3.225]$$

where λ is a constant such that $0 < \lambda < E_0$.

(1) *Matrix representing the Hamiltonian*

In the orthonormal set $\{|\Phi_n\rangle\}$, the Hamiltonian $H = H_0 + W$ is represented by the square matrix 3 × 3 of matrix elements:

$$H_{nm} = H_{0nm} + W_{nm} \qquad [3.226]$$

This yields:

$$(H) = \begin{pmatrix} H_{011}+W_{11} & H_{012}+W_{12} & H_{013}+W_{13} \\ H_{021}+W_{21} & H_{022}+W_{22} & H_{023}+W_{23} \\ H_{031}+W_{31} & H_{032}+W_{32} & H_{033}+W_{33} \end{pmatrix} \qquad [3.227]$$

Using [3.224], we find:

$$H_{0nm} = \langle \Phi_n | H_0 | \Phi_m \rangle = H_0 \langle \Phi_n | \Phi_m \rangle \Rightarrow H_{0nm} = E_0 \delta_{nm} \qquad [3.228]$$

Using [3.225] and [3.228], matrix [3.227] is finally written as:

$$(H) = \begin{pmatrix} E_0 & \frac{\lambda}{\sqrt{2}} & \frac{\lambda}{\sqrt{2}} \\ \frac{\lambda}{\sqrt{2}} & E_0 & 0 \\ \frac{\lambda}{\sqrt{2}} & 0 & E_0 \end{pmatrix} \qquad [3.229]$$

(2) *Expressions of the eigenvalues and of the eigenstates*

The eigenkets $|\phi_-\rangle$, $|\phi_0\rangle$ and $|\phi_+\rangle$ of H are associated with the eigenvalues α_-, α_0 and α_+, respectively.

(2.1) *Expressions of the eigenvalues*

The eigenvalues α_-, α_0 and α_+ are determined by the characteristic equation:

$$Det\,[(H) - \alpha I] = 0 \qquad [3.230]$$

where (H) designates the matrix [3.229], α designates α_-, α_0 or α_+ and I is the unit matrix. Then we obtain:

$$\text{Det}\begin{pmatrix} E_0-\alpha & \frac{\lambda}{\sqrt{2}} & \frac{\lambda}{\sqrt{2}} \\ \frac{\lambda}{\sqrt{2}} & E_0-\alpha & 0 \\ \frac{\lambda}{\sqrt{2}} & 0 & E_0-\alpha \end{pmatrix}=0 \qquad [3.231]$$

This leads to:

$$(E_0 - \alpha)[(E_0 - \alpha)^2 - \lambda^2] = 0 \qquad [3.232]$$

The eigenvalues of the Hamiltonian H can be deduced from the equation:

$$\alpha_0 = E_0;\ \alpha_- = E_0 - \lambda;\ \alpha_+ = E_0 + \lambda \qquad [3.233]$$

(2.2) *Expressions of the eigenstates*

We write the equations with eigenvalues of the Hamiltonian H. We get:

$$\begin{cases} H|\phi_-\rangle = \alpha_-|\phi_-\rangle \\ H|\phi_0\rangle = \alpha_0|\phi_0\rangle \\ H|\phi_+\rangle = \alpha_+|\phi_+\rangle \end{cases} \qquad [3.234]$$

In the basis $\{|\Phi_n\rangle\}$ forming the eigenstates of H_0, the eigenstates of H satisfy the expansion:

$$|\phi_\nu\rangle = \sum_{n=1}^{3} c_{n\nu}|\Phi_n\rangle \qquad [3.235]$$

Then we obtain:

$$\begin{cases} |\phi_-\rangle = c_{1-}|\Phi_1\rangle + c_{2-}|\Phi_2\rangle + c_{3-}|\Phi_3\rangle \\ |\phi_0\rangle = c_{10}|\Phi_1\rangle + c_{20}|\Phi_2\rangle + c_{30}|\Phi_3\rangle \\ |\phi_+\rangle = c_{1+}|\Phi_1\rangle + c_{2+}|\Phi_2\rangle + c_{3+}|\Phi_3\rangle \end{cases} \qquad [3.236]$$

Let us summarize the system of equations [3.234] as follows:

$$H|\phi_v\rangle = \alpha_v|\phi_v\rangle \quad [3.237]$$

Equation [3.237] can be written using [3.229] and [3.235] in matrix form:

$$\begin{pmatrix} E_0 & \frac{\lambda}{\sqrt{2}} & \frac{\lambda}{\sqrt{2}} \\ \frac{\lambda}{\sqrt{2}} & E_0 & 0 \\ \frac{\lambda}{\sqrt{2}} & 0 & E_0 \end{pmatrix} \begin{pmatrix} c_{1v} \\ c_{2v} \\ c_{3v} \end{pmatrix} = \alpha_v \begin{pmatrix} c_{1v} \\ c_{2v} \\ c_{3v} \end{pmatrix} \quad [3.238]$$

From equation [3.238], we deduce the following system:

$$\begin{cases} E_0 c_{1v} \quad \frac{\lambda}{\sqrt{2}}(c_{2v}+c_{3v}) = \alpha_v c_{1v} & (1) \\ \frac{\lambda}{\sqrt{2}} c_{1v} + E_0 c_{2v} = \alpha_v c_{2v} & (2) \\ \frac{\lambda}{\sqrt{2}} c_{1v} + E_0 c_{3v} = \alpha_v c_{3v} & (3) \end{cases} \quad [3.239]$$

The coefficients c_{nv} in [3.239] are fully determined taking the normalization condition into consideration:

$$|c_{1v}|^2 + |c_{2v}|^2 + |c_{3v}|^2 = 1 \quad [3.240]$$

Making the sum of equations (2) and (3) of the system [3.239] and deducting (3) from equation (2), we get:

$$\begin{cases} \sqrt{2}\lambda c_{1v} = (c_{2v}+c_{3v})(\alpha_v - E_0) & (4) \\ \alpha_v(c_{2v}-c_{3v}) = E_0(c_{2v}-c_{3v}) & (5) \end{cases} \quad [3.241]$$

Equation (5) of system [3.241] has two solutions: $\alpha_v = E_0$ and $\alpha_v \neq E_0$.

– *For* $\alpha_\nu = E_0 = \alpha_0$

Equations (1) and (4) of systems [3.239] and [3.241] yield:

$$\begin{cases} c_{10} = 0 \\ c_{20} = -c_{30} \end{cases} \qquad [3.242]$$

Using [3.240] and [3.242], we find:

$$c_{20} = -c_{30} = \frac{1}{\sqrt{2}} \qquad [3.243]$$

The eigenstate $|\phi_0\rangle$ of H is then written according to [3.235]:

$$|\phi_0\rangle = \frac{1}{\sqrt{2}}|\Phi_1\rangle - \frac{1}{\sqrt{2}}|\Phi_2\rangle \qquad [3.244]$$

– *For* $\alpha_\nu \neq E_0 \, (\nu \neq 0)$

Equations (5) and (4) of [3.241] yield, respectively:

$$\begin{cases} c_{2\nu} = c_{3\nu} \\ \lambda c_{1\nu} = \sqrt{2} c_{2\nu}(\alpha_\nu - E_0) \end{cases} \qquad [3.245]$$

– *For* $\alpha_\nu = \alpha_- = E_0 - \lambda$

Using [3.245], we get:

$$\begin{cases} c_{2-} = c_{3-} \\ c_{1-} = -\sqrt{2}\, c_{2-} \end{cases} \qquad [3.246]$$

Similarly, taking [3.240] into account, we find:

$$c_{2-} = \frac{1}{\sqrt{2}} ; c_{2-} = c_{3-} = -\frac{1}{2} \qquad [3.247]$$

Using the expansion [3.235], the eigenstate $|\phi_-\rangle$ of H is written as:

$$|\phi_-\rangle = \frac{1}{\sqrt{2}}|\Phi_1\rangle - \frac{1}{2}|\Phi_2\rangle - \frac{1}{2}|\Phi_3\rangle \qquad [3.248]$$

– *For* $\alpha_\nu = \alpha_+ = E_0 + \lambda$

Considering the system [6.402], we get:

$$\begin{cases} c_{2+} = c_{3+} \\ c_{1+} = \sqrt{2}\, c_{2+} \end{cases} \qquad [3.249]$$

According to [3.240], this leads to:

$$c_{1+} = \frac{1}{\sqrt{2}}; c_{2+} = c_{3+} = \frac{1}{2} \qquad [3.250]$$

The eigenstate $|\phi_+\rangle$ of H is then written as:

$$|\phi_+\rangle = \frac{1}{\sqrt{2}}|\Phi_1\rangle + \frac{1}{2}|\Phi_2\rangle + \frac{1}{2}|\Phi_3\rangle \qquad [3.251]$$

In summary, the eigenstates of the Hamiltonian H are written as:

$$\begin{cases} |\phi_-\rangle = \frac{1}{\sqrt{2}}|\Phi_1\rangle - \frac{1}{2}|\Phi_2\rangle - \frac{1}{2}|\Phi_3\rangle \\ |\phi_0\rangle = \frac{1}{\sqrt{2}}|\Phi_2\rangle - \frac{1}{\sqrt{2}}|\Phi_3\rangle \\ |\phi_+\rangle = \frac{1}{\sqrt{2}}|\Phi_1\rangle + \frac{1}{2}|\Phi_2\rangle + \frac{1}{2}|\Phi_3\rangle \end{cases} \qquad [3.252]$$

(3.1) *State of the system*

We express the initial state $|\psi(0)\rangle = |\Phi_1\rangle$ of the system in the basis of eigenstates of the Hamiltonian H. Let us add the first and third of the expansions of system [3.352]. We obtain:

$$|\psi(0)\rangle = \frac{1}{\sqrt{2}}|\phi_-\rangle - \frac{1}{\sqrt{2}}|\phi_+\rangle \qquad [3.253]$$

The state of the system at instant t is then written as:

$$|\psi(t)\rangle = \sum_{\nu=-1}^{+1} \sum_{n=1}^{2} c_{n\nu}(0) e^{-i\frac{\alpha_\nu}{\hbar}t} |\phi_\nu\rangle \qquad [3.254]$$

which is:

$$|\psi(t)\rangle = c_{1-}(0)\,e^{-i\alpha_- t/\hbar}|\phi_-\rangle + c_{2+}(0)\,e^{-i\alpha_+ t/\hbar}|\phi_+\rangle \quad [3.255]$$

Taking [3.233], [3.247] and [3.250] into account, we have:

$$|\psi(t)\rangle = \frac{1}{\sqrt{2}}\left\{e^{-i(E_0-\lambda)t/\hbar}|\phi_-\rangle + e^{-i(E_0-\lambda)t/\hbar}|\phi_+\rangle\right\}$$

or after factorization:

$$|\psi(t)\rangle = \frac{1}{\sqrt{2}}e^{-iE_0 t/\hbar}\left\{e^{i\lambda t/\hbar}|\phi_-\rangle + e^{-i\lambda t/\hbar}|\phi_+\rangle\right\} \quad [3.256]$$

(3.2) *Calculation of the mean value*

The mean value $\langle H\rangle$ (t) of the observable H is given by the relation:

$$\langle H\rangle(t) = \langle\psi(t)|H|\psi(t)\rangle \quad [3.257]$$

Using [3.256], we get:

$$H|\psi(t)\rangle = \frac{1}{\sqrt{2}}e^{-iE_0 t/\hbar}\left\{e^{i\lambda t/\hbar}H|\phi_-\rangle + e^{-i\lambda t/\hbar}H|\phi_+\rangle\right\} \quad [3.258]$$

Hence:

$$H|\psi(t)\rangle = \frac{1}{\sqrt{2}}e^{-iE_0 t/\hbar}\left\{e^{i\lambda t/\hbar}\alpha_-|\phi_-\rangle + e^{-i\lambda t/\hbar}\alpha_+|\phi_+\rangle\right\} \quad [3.259]$$

Moreover, the bra corresponding to ket [3.256] is written as:

$$\langle\psi(t)| = \frac{1}{\sqrt{2}}e^{iE_0 t/\hbar}\left\{e^{-i\lambda t/\hbar}\langle\phi_-| + e^{i\lambda t/\hbar}\langle\phi_+|\right\} \quad [3.260]$$

Using [3.259] and [3.260], the mean value [3.257] is written as:

$$\langle H\rangle(t) = \frac{1}{2}(\alpha_- + \alpha_+) \qquad [3.261]$$

Using [3.233], relation [3.261] is written as:

$$\langle H\rangle(t) = \frac{1}{2}[(E_0 - \lambda) + (E_0 + \lambda)]$$

or finally:

$$\langle H\rangle(t) = E_0 \qquad [3.262]$$

CONCLUSION.– The mean value $\langle H\rangle(t)$ does not depend on time. This is due to the fact that the system is conservative.

3.5.11. *Solution 11 – Evolution of the density operator*

The state vector is given by the expansion:

$$|\psi(t)\rangle = \sum_n c_n |u_n\rangle \qquad [3.263]$$

The density operator is defined by the relation:

$$\rho(t) = |\psi(t)\rangle\langle\psi(t)| \qquad [3.264]$$

(1) *Trace, mean value*

– *Trace of ρ (t)*

Using [3.264], the trace of the density operator is written as:

$$Tr\rho(t) = \sum_n \langle u_n|\psi(t)\rangle\langle u_n|\psi(t)\rangle$$

Hence:

$$Tr\rho(t) = \sum_n |\langle u_n|\psi(t)\rangle|^2 = \sum_n |c_n|^2 = 1 \qquad [3.265]$$

– *Mean value of A*

The state vector [3.263] being normed, the mean value $\langle A\rangle(t)$ of the observable A is given by the relation:

$$\langle A\rangle(t) = \langle\psi(t)|A|\psi(t)\rangle \qquad [3.266]$$

We insert in [3.266] the closing relation satisfied by the discrete set $\{|u_n\rangle\}$. We have:

$$\langle A\rangle(t) = \sum_n \langle\psi(t)|A|u_n\rangle\langle u_n|\psi(t)\rangle$$

Hence:

$$\langle A\rangle(t) = \sum_n \langle u_n|\psi(t)\rangle\langle\psi(t)|A|u_n\rangle \qquad [3.267]$$

Taking [3.264] into account, relation [3.267] is written as:

$$\langle A\rangle(t) = \sum_n \langle u_n|\rho(t)A|u_n\rangle$$

or finally:

$$\langle A\rangle(t) = Tr[\rho(t)A] \qquad [3.268]$$

(2) *Evolution equation*

Let us differentiate [3.264] with respect to time. We get:

$$\frac{d\rho(t)}{dt} = \left[\frac{d}{dt}|\psi(t)\rangle\right]\langle\psi(t)| + |\psi(t)\rangle\left[\frac{d}{dt}\langle\psi(t)|\right] \qquad [3.269]$$

Moreover, Schrödinger's equation describing the temporal evolution of the system is given by the relation:

$$i\hbar\frac{d}{dt}|\psi(t)\rangle = H(t)|\psi(t)\rangle \qquad [3.270]$$

Knowing that H is Hermitian, the complex conjugate of equation [3.270] is written as:

$$-i\hbar\frac{d}{dt}\langle\psi(t)| = \langle\psi(t)|H(t) \quad [3.271]$$

Using [3.270] and [3.271], the evolution equation [3.269] is written as follows:

$$\frac{d\rho(t)}{dt} = \frac{1}{i\hbar}H(t)|\psi(t)\rangle - \frac{1}{i\hbar}|\psi(t)\rangle\langle\psi(t)|H(t)$$

which is:

$$\frac{d\rho(t)}{dt} = \frac{1}{i\hbar}\{H(t)\rho(t) - \rho(t)H(t)\} \quad [3.272]$$

The commutator of operators H and ρ can be identified in [3.272]. Finally:

$$\frac{d\rho(t)}{dt} = \frac{1}{i\hbar}[H(t), \rho(t)] \quad [3.273]$$

(3) *Proof*

Using [3.273], we get:

$$\langle u_n|\frac{d\rho(t)}{dt}|u_k\rangle = \frac{1}{i\hbar}\langle u_n|[H(t), \rho(t)]|u_k\rangle$$

Hence:

$$i\hbar\frac{d}{dt}\langle u_n|\rho(t)|u_k\rangle = \langle u_n|H(t)\rho(t)|u_k\rangle - \langle u_n|\rho(t)H(t)|u_k\rangle \quad [3.274]$$

Using the equation with eigenvalues of the Hamiltonian H and taking its hermiticity into account, the equation gives:

$$i\hbar\frac{d}{dt}\langle u_n|\rho(t)|u_k\rangle = E_n\langle u_n|\rho(t)|u_k\rangle - E_k\langle u_n|\rho(t)|u_k\rangle$$

Finally, the evolution equation of the density operator is written as follows:

$$i\hbar\frac{d}{dt}\rho_{n,k}(t) = (E_n - E_k)\rho_{n,k}(t) \quad [3.275]$$

(4) *Proof*

By definition, populations are the elements of matrix $\rho_{n,n}(t)$. For $n = k$, the evolution equation [3.275] gives:

$$\rho_{n,n} = \text{Cst} \qquad [3.276]$$

Populations are therefore constant.

By definition, coherences are the elements of matrix $\rho_{n,k}(t)$ for $n \neq k$. Integrating [3.275], we get:

$$\rho_{n,k}(t) = \rho_{n,k}(0)e^{-\frac{(E_n - E_k)}{\hbar}t} \qquad [3.277]$$

Result [3.277] shows that coherences oscillate with the single Bohr frequency given by $\nu_{n,k} = (E_n - E_k)/h$.

3.5.12. *Solution 12 – Evolution of a 1/2 spin in a magnetic field*

We summarize the essential given data:

– Hamiltonian H describing the spin evolution in the magnetic field:

$$H = \omega S_z \qquad [3.278]$$

with $\omega = -2\gamma B$

– Equation with eigenvalues of the Hamiltonian H:

$$\begin{aligned} H|+\rangle &= E^{+}|+\rangle = \frac{\hbar\omega}{2}|+\rangle \\ H|-\rangle &= E^{-}|-\rangle = \frac{\hbar\omega}{2}|-\rangle \end{aligned} \qquad [3.279]$$

– State of the spin at $t = 0$:

$$|\psi(0)\rangle = \cos\frac{\theta}{2}e^{-i\frac{\varphi}{2}}|+\rangle + \sin\frac{\theta}{2}e^{i\frac{\varphi}{2}}|-\rangle \qquad [3.280]$$

(1) *Verification*

Relation [3.278] shows that H and S_z commute ($HS_z - S_zH = 0$). Consequently, the physical quantities associated with them are simultaneously measurable.

(2) *Calculation of mean values, conclusion*

Mean value is given as:

$$\langle H \rangle_0 = \langle \psi(0) | H | \psi(0) \rangle \qquad [3.281]$$

Using [3.280], we have:

$$H|\psi(0)\rangle = \cos\frac{\theta}{2} e^{-i\frac{\varphi}{2}} E^+ |+\rangle + \sin\frac{\theta}{2} e^{i\frac{\varphi}{2}} E^- |-\rangle \qquad [3.282]$$

Moreover, the bra corresponding to ket [3.280] is written as:

$$\langle \psi(0)| = \cos\frac{\theta}{2} e^{i\frac{\varphi}{2}} \langle +| + \sin\frac{\theta}{2} e^{-i\frac{\varphi}{2}} \langle -| \qquad [3.283]$$

Using [3.282] and [3.283], the mean value [3.281] is written as:

$$\langle H \rangle_0 = E^+ \cos^2\frac{\theta}{2} + E^- \sin^2\frac{\theta}{2} \qquad [3.284]$$

Using [3.279], relation [3.284] gives:

$$\langle H \rangle_0 = \frac{\hbar\omega}{2}\left(\cos^2\frac{\theta}{2} - \sin^2\frac{\theta}{2} \right)$$

or finally:

$$\langle H \rangle_0 = \frac{\hbar\omega}{2} \cos\theta \qquad [3.285]$$

We have used the trigonometric transformation:

$$\cos\left(\frac{\theta}{2} + \frac{\theta}{2} \right) = \left(\cos^2\frac{\theta}{2} - \sin^2\frac{\theta}{2} \right)$$

The mean value $\langle S_z \rangle_0$ is deduced from [3.278], taking [3.285] into account. Hence:

$$\langle S_z \rangle_0 = \frac{1}{\omega} \langle H \rangle_0 = \frac{\hbar}{2} \cos\theta \qquad [3.286]$$

CONCLUSION.– The angle θ being constant, results [3.285] and [3.286] show that the observables H and S_z are constants of motion.

(3) *Study of the system at an instant t*

(3.1) *Expression of the state vector*

The state vector at instant t is given by the expression:

$$|\psi(t)\rangle = \sum_{n=1}^{2} c_n(0) e^{-i\frac{E_n}{\hbar}t} |\phi_n\rangle \tag{3.287}$$

Hence:

$$|\psi(t)\rangle = c_1(0) e^{-i\frac{E_1}{\hbar}t} |\phi_1\rangle + c_2(0) e^{-i\frac{E_2}{\hbar}t} |\phi_2\rangle \tag{3.288}$$

At $t = 0$, we have:

$$|\psi(0)\rangle = c_1(0)|\phi_1\rangle + c_2(0)|\phi_2\rangle \tag{3.289}$$

Comparing [3.289] and [3.280], we can see that:

$$\begin{cases} c_1(0) = \cos\frac{\theta}{2} e^{-i\frac{\varphi}{2}};\ c_2(0) = \sin\frac{\theta}{2} e^{i\frac{\varphi}{2}} \\ |\phi_1\rangle = |+\rangle ; |\phi_2\rangle = |-\rangle \end{cases} \tag{3.290}$$

Taking into account the evolution operator [3.55] and using [3.290], the state vector [3.288] is written as:

$$|\psi(t)\rangle = \cos\frac{\theta}{2} e^{-i\frac{\varphi}{2}} e^{-i\frac{H}{\hbar}t} |+\rangle + \sin\frac{\theta}{2} e^{i\frac{\varphi}{2}} e^{-i\frac{H}{\hbar}t} |-\rangle$$

That means:

$$|\psi(t)\rangle = \cos\frac{\theta}{2} e^{-i\frac{\varphi}{2}} e^{-i\frac{E^+}{\hbar}t} |+\rangle + \sin\frac{\theta}{2} e^{i\frac{\varphi}{2}} e^{-i\frac{E^-}{\hbar}t} |-\rangle \tag{3.291}$$

or, using [3.279]:

$$|\psi(t)\rangle = \cos\frac{\theta}{2}e^{-i\frac{(\omega t+\varphi)}{2}}|+\rangle + \sin\frac{\theta}{2}e^{i\frac{(\omega t+\varphi)}{2}}|-\rangle \qquad [3.292]$$

The evolution equation [3.292] shows that the presence of the magnetic field introduces a phase difference that is proportional to the time between the coefficients assigned to the eigenvectors $|\pm\rangle$ of the observable of spin S_z.

(3.2) *Finding the mean values, conclusion*

Let us express the mean value $\langle H\rangle(t)$:

$$\langle H\rangle(t) = \langle\psi(t)|H|\psi(t)\rangle = \langle\psi(0)|H|\psi(0)\rangle \qquad [3.293]$$

Using [3.293], we find as previously:

$$\langle H\rangle(t) = \langle H\rangle_0 = E^+\cos^2\frac{\theta}{2} + E^-\sin^2\frac{\theta}{2} = \frac{\hbar\omega}{2}\cos\theta \qquad [3.294]$$

and:

$$\langle S_z\rangle(t) = \langle S\rangle_0 = \frac{\hbar}{2}\cos\theta \qquad [3.295]$$

(3.3) *Simultaneous measurement*

The simultaneous measurement of the energy E and of the components associated with observables S_x and S_y is impossible since H, S_x and S_y do not form a CSCO. Indeed, the observables S_z, S_x and S_y satisfy the commutation relations:

$$[S_x, S_y] = i\hbar S_z \;;\; [S_y, S_z] = i\hbar S_x \;;\; [S_z, S_x] = i\hbar S_y \qquad [3.296]$$

Since $H = \omega S_z$, it follows that H commutes with neither S_x nor S_y.

(3.4) *Calculation of mean values*

Matrices representing the observables S_x and S_y in the basis $\{|+\rangle \text{ and } |-\rangle\}$:

$$(S_x) = \frac{\hbar}{2}\begin{pmatrix}0 & 1\\ 1 & 0\end{pmatrix} \;;\; (S_y) = \frac{\hbar}{2}\begin{pmatrix}0 & -i\\ i & 0\end{pmatrix} \qquad [3.297]$$

The mean values $\langle S_x\rangle(t)$ and $\langle S_y\rangle(t)$ are written as, respectively:

$$\langle S_x\rangle(t)=\langle\psi(t)|S_x|\psi(t)\rangle\,;\langle S_y\rangle(t)=\langle\psi(t)|S_y|\psi(t)\rangle \qquad [3.298]$$

Knowing that a ket is represented by a single column matrix and a bra by a single line matrix, using [3.280] and [3.283], expressions [3.298] are written as:

$$\langle S_x\rangle(t)=\frac{\hbar}{2}\begin{pmatrix}\cos\frac{\theta}{2}e^{i\frac{(\omega t+\varphi)}{2}} & \sin\frac{\theta}{2}e^{-i\frac{(\omega t+\varphi)}{2}}\end{pmatrix}\begin{pmatrix}0 & 1\\ 1 & 0\end{pmatrix}\begin{pmatrix}\cos\frac{\theta}{2}e^{-i\frac{(\omega t+\varphi)}{2}}\\ \sin\frac{\theta}{2}e^{i\frac{(\omega t+\varphi)}{2}}\end{pmatrix} \qquad [3.299]$$

and:

$$\langle S_x\rangle(t)=\frac{\hbar}{2}\begin{pmatrix}\cos\frac{\theta}{2}e^{i\frac{(\omega t+\varphi)}{2}} & \sin\frac{\theta}{2}e^{-i\frac{(\omega t+\varphi)}{2}}\end{pmatrix}\begin{pmatrix}0 & -i\\ i & 0\end{pmatrix}\begin{pmatrix}\cos\frac{\theta}{2}e^{-i\frac{(\omega t+\varphi)}{2}}\\ \sin\frac{\theta}{2}e^{i\frac{(\omega t+\varphi)}{2}}\end{pmatrix} \qquad [3.300]$$

We finally obtain:

$$\langle S_x\rangle(t)=\frac{\hbar}{2}\sin\theta\cos(\omega t+\varphi);\langle S_y\rangle(t)=\frac{\hbar}{2}\sin\theta\sin(\omega t+\varphi) \qquad [3.301]$$

CONCLUSION.– The mean values $\langle S_x\rangle$ (t) and $\langle S_y\rangle$ (t) oscillate in time with the single Bohr frequency $\nu = \omega/2\pi$. This explains the fact that observables S_x and S_y are not constants of motion.

Appendix 1

Quantum Well of Semiconductor Materials

A1.1. 2D, 1D and 0D confinement

At temperature $T = 0$ K, the energy bands of a crystal are filled with valence electrons. The last of these bands is known as a *valence band* (VB) of energy E_v separated from a *conduction band* (CB) of energy E_c by a *forbidden band* (FB) corresponding to the *material gap* of positive energy $E_g = E_c - E_v$ (Figure A1.1). In these energy bands, Schrödinger's equation has no solution.

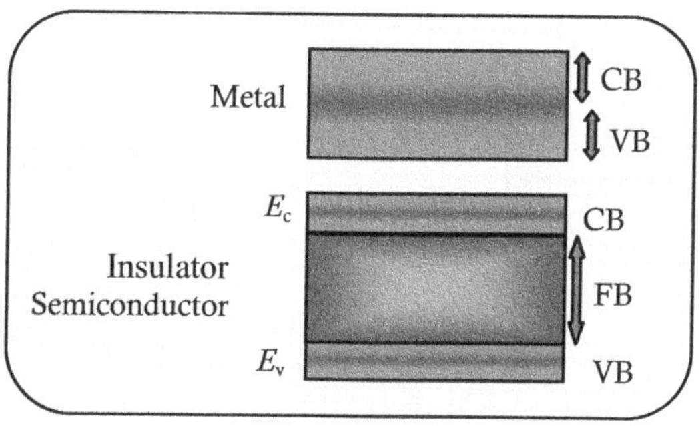

Figure A1.1. *Relative arrangement of the energy bands in a material*

Three categories of materials can be distinguished, according to the gap value: *metals*, *insulators* and *semiconductors*. As an example, let us consider several values of the gap at 300 K [SAK 15]. Metal: tin (Sn): 0.0 eV; insulator: diamond (C): 6.0 eV; semiconductor: silicon (Si): 1.12 eV; germanium: 0.67 eV; gallium arsenide (GaAs): 1.40 eV.

– Metals have low resistivity at ambient temperature (of about 10^{-5} Ωcm). Conduction is due to the free electrons in the CB (density: 10^{22} to 10^{28} cm^{-3}). An increase in temperature drives only a small increase in resistivity, because the motion of free electrons is hindered by the vibrations of metal atoms.

– The resistivity of insulators is above 10^8 Ω cm. This is the case for glass, mica, silica (SiO_2), carbon, etc. In insulators, the release of electrons is triggered by an increase in temperature. This drives a decrease in resistivity.

– Semiconductors are materials whose resistivity ranges between 10^{-8} and 10^4 Ω cm (or more). Conduction is ensured by the charge carriers (electrons and holes). Two types of semiconductors can be distinguished: *intrinsic* semiconductors (Si, Ge, As, etc.) and *extrinsic* semiconductors (GaAs, GaAlAs, etc.).

In a *perfect three-dimensional crystal* (3D), energy is not discrete. When the charge carriers are *confined* in a sufficiently low volume (of about *De Broglie* wavelength ($\lambda = h/p$) of the charge carrier), *discretization* of the energy values can be observed. The *nanocrystal* is then described in a purely quantum context. Three types of *nanostructures* can be identified [SAK 15]:

– *Quantum well* (material deposited in a 2D layer): an *electron confinement* is introduced along a direction (*Oz*, for example). This is known as *2D confinement* (Figure A1.2). A discretization of the energy levels (E_{n_z}) is associated with this confinement. In the *2D* confinement, electrons can freely move along two directions.

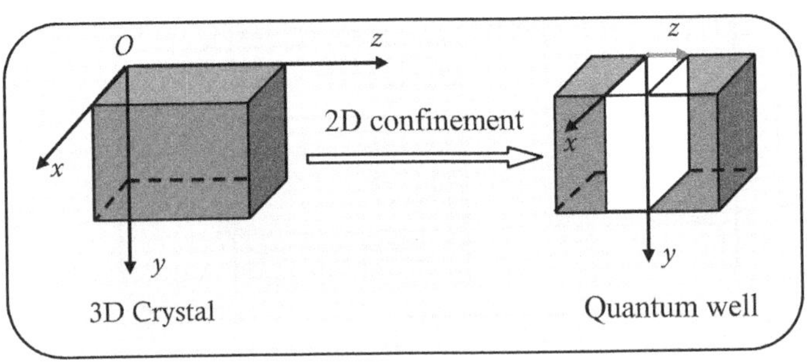

Figure A1.2. *Quantum well (2D confinement)*

– *Quantum wire* (material deposited in a 1D layer): An electron confinement is introduced along two directions (*Ox* and *Oy*, for example). This is known as *1D confinement* (Figure A1.3).

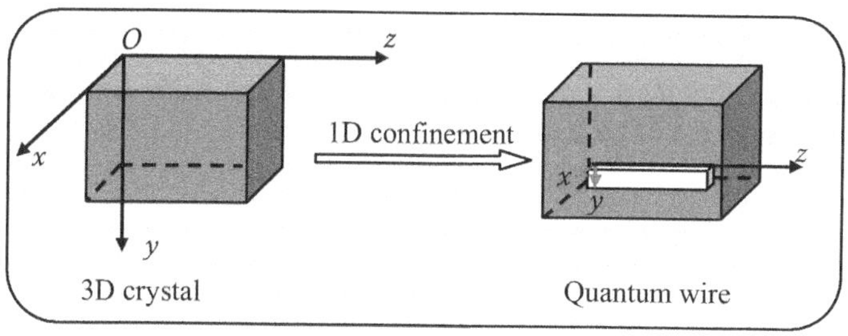

Figure A1.3. *Quantum wire (1D confinement)*

A discretization of the energy levels ($E_{nx,ny}$) is associated with this confinement. In this 1D confinement, electrons are free to move along a single direction (*Oz*).

– *Quantum dot* (material deposited in 0D layer): An electron confinement is introduced along three directions (*Ox*, *Oy* and *Oz*). This is a *0D confinement* (Figure A1.4). A *total discretization* of the energy levels ($E_{nx,ny,nz}$) is associated with this type of confinement. In a 0D confinement, electrons are not free to move along any direction: the nanostructure is assimilated to a *quantum dot.*

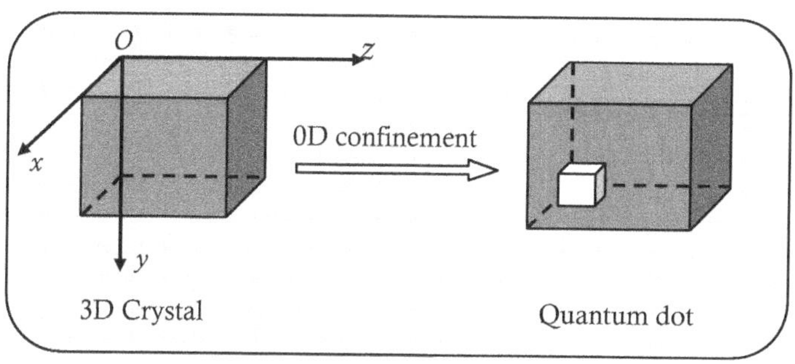

Figure A1.4. *Quantum dot (0D confinement)*

There are few systems that can be assimilated to model quantum wires (these are wires that are to little or no extent perturbed by disorder). Such an example are *carbon nanotubes*; semiconductor wires obtained by sophisticated techniques of *nanostructuring* by *epitaxy*. Epitaxy is the technique that enables crystal growth on a crystal. *Polydiacetylene* (known as poly-3BCMU) is an example of ideal *quantum wire* having the perfect structure of a carbon nanotube [DUB 04, BAR 06]. The chains of poly-3BCMU are obtained by triggering the polymerization of the *diacetylene monomer* by exposure to various radiations (UV, slow electrons or gamma radiations). The reaction takes place in solid phase. The average length of chains easily exceeds a dozen microns (over 20,000 chained monomer units). The following sections focus on a brief study of the quantum wells and quantum dots of semiconductor materials as a connection to the quantum wells and dots studied in sections 1.6.1 and 1.6.4 of Chapter 1.

A1.2. Description of the quantum wells of semiconductors

In semiconductors, a quantum well can be obtained by the successive stacking of plane layers of various materials [BAR 06]. For example, a *well material* (B) of a small gap surrounded on each side by a *barrier material* (A) of a larger gap (Figure A1.5). In this case, electrons see a square potential well with a finite height barrier.

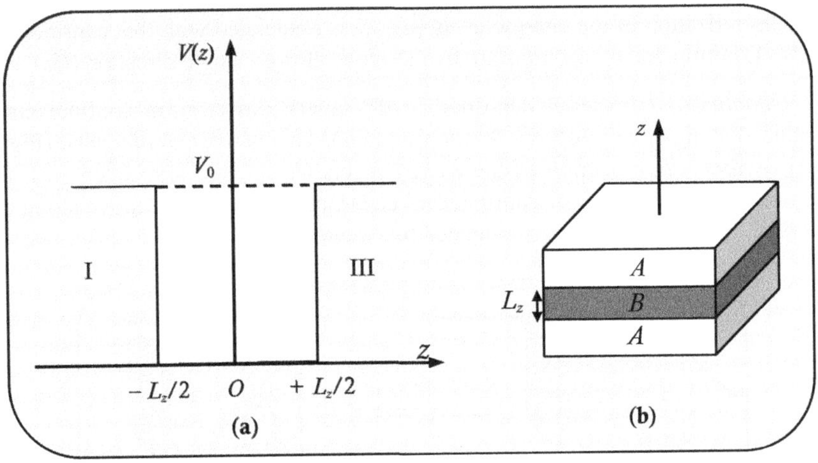

Figure A1.5. *Quantum well: (a) quantum well of width Lz and height V0. (b) Quantum well constituted of a layer of width Lz of semiconductor material (B) of a small gap surrounded on each side by a layer of semiconductor material (A) of a larger gap*

The development of quantum wells has greatly benefitted from the progress achieved in the growth of materials. Quantum wells are mainly manufactured by *molecular beam epitaxy* or by *chemical plating* in vapor phase. The discontinuity between energy bands in the two materials creates a *potential barrier* that confines the charge carriers (electrons and holes) in the quantum well. Depending on the nature of the discontinuity of the FBs at the *well-barrier interface*, three types of quantum wells can be identified [LAH 17]:

1) *quantum well of type I*: electrons and holes are confined in the same material constituting the well. The energy of the emitted photon depends on the confinement energies of the carriers and of the well material gap;

2) *quantum well of type II,* in which electrons and holes are confined in the same material constituting the well. The energy of the emitted photon depends on the confinement energies of the carriers and on the well material gap;

3) *quantum well of type III,* which is a particular case of type II; the lower level of the CB is below the highest level of the VB. In this configuration, known as *semimetallic*, the energy of the emitted photon is only a function of the confinement energies of electrons and holes.

A1.3. Wave function and levels of energy

In a semiconductor, the moving mass of the charge carriers is the "*effective mass*" denoted by m^*, which differs from the rest mass of the electron m_0. If $m^*_c = m_c$ designates the mass of an electron in the CB and $m^*_v = m_v$ is the mass of a hole in the VB, then [SAK 15]:

$$m_c = 1.05\ m_0;\ m_v = 0.62\ m_0;\ m_0 = 9.1 \times 10^{-28}\ \text{g}. \qquad \text{[A1.1]}$$

Unless otherwise specified, we put: $m_c = m_v = m_0$.

Let us then consider the quantum well of height V_0 and width L_z (Figure A1.5(a)). The potential $V(z)$ is defined by:

$$V(z) = \begin{cases} 0 & if\ |z| < L_z/2 \\ V_0 & if\ |z| > L_z/2 \end{cases} \qquad \text{[A1.2]}$$

In the absence of electron-hole interaction, Schrödinger's equation is ($m = m^*$):

$$\left[-\frac{\hbar^2}{2m} \nabla + V(z) \right] \Psi(x,y,z) = E\,\Psi(x,y,z) \qquad \text{[A1.3]}$$

As the potential $V(z)$ is a function of the single variable z, the wave function can be built as the product of a function $\Phi\ (x, y)$ describing the motion of the particle in the xy plane of the quantum well and a function $\chi\ (z)$ for the motion along direction z, hence:

$$\Psi\ (x, y, z) = \Phi\ (x, y) \times \chi\ (z) \qquad \text{[A1.4]}$$

The energy E of the particle is the sum of two terms and is written as follows:

$$E = E_z^n + \frac{\hbar^2}{2m}k_x^2 + \frac{\hbar^2}{2m}k_y^2 \qquad [A1.5]$$

The first term in the right-hand side of equation [A1.5] designates the confinement energy and the second and the third terms designate the sum of the kinetic energies of the particle in the xy plane. The exponent n represents the quantum number characterizing the discrete character of the energy along the confinement direction Oz. Considering [A1.4], the resolution of Schrödinger's equation [A1.3] gives the following solutions for the wave function χ (z) (see exercise in section 1.7.7):

– if n is even:

$$\chi(z) = \begin{cases} A\cos k_z z & if\ |z| < L_z/2 \\ Ce^{\pm\kappa z} & if\ |z| > L_z/2 \end{cases} \qquad [A1.6]$$

– if n is odd:

$$\chi(z) = \begin{cases} B\sin k_z z & if\ |z| < L_z/2 \\ \pm Ce^{\pm\kappa z} & if\ |z| > L_z/2 \end{cases} \qquad [A1.7]$$

In equations [A1.6] and [A1.7]:

$$k_z^2 = \frac{2m_w^* E_z^n}{\hbar^2}\ ;\ \kappa^2 = \frac{2m_b^*(V_0 - E_z^n)}{\hbar^2} \qquad [A1.8]$$

In these expressions:

– m_w* is the mass of the particle in the quantum well material;

– m_b* is the mass of the particle in the barrier material.

It is worth noting that these expressions are similar to relations [1.244], reproduced below:

$$\rho^2 = -\frac{2mE}{\hbar^2}\ ;\ k^2 = \frac{2m(E+V_0)}{\hbar^2}$$

It can be seen that in the case of a potential well of a semiconductor material, the mass m in expressions [1.244] must be indexed so that it takes into account the state of the particle in the barrier material and in the well material, as indicated by relations [A1.8]. The confinement energy E_z^n is the solution to equations:

$$\tan\left[\sqrt{\frac{m_\omega^* E_z^n}{2\hbar^2}} L_z\right] = \sqrt{\frac{V_0 - E_z^n}{E_z^n}}\text{, if } n \text{ is even} \quad \text{[A1.9]}$$

$$\cot\left[\sqrt{\frac{m_\omega^* E_z^n}{2\hbar^2}} L_z\right] = -\sqrt{\frac{V_0 - E_z^n}{E_z^n}}\text{, if } n \text{ is odd} \quad \text{[A1.10]}$$

The graphical resolution of equations [A1.41] and [A1.42] is similar to the resolution of equations [1.267] and [1.268]. The resulting curves are similar to those represented in Figure 1.21.

For a quantum well of depth $V_0 \approx \infty$, the particle is confined in the well. Schrödinger's equation resolution in zone II yields the energy:

$$E_z^n = \frac{\hbar^2 \pi^2}{2 m_\omega^* L_z^2} n^2 \quad \text{[A1.11]}$$

This leads to expression [1.63] that is reproduced below (putting $m_\omega^* = m$ and $L_z = a$):

$$E_n = \frac{\hbar^2 \pi^2}{2ma^2} n^2$$

Appendix 2

Quantum Dot of Semiconductor Materials

A2.1. Definition, qubits

A *quantum dot* is a *heterostructure* obtained by 0D confinement of semiconductor materials. Current research on the properties of quantum dots is motivated by the fact that they constitute potential elementary bricks of the *quantum processing of information*, known as *quantum bits* or *qubits*. On a long term, it may be possible to use the properties of quantum dots at a nanometric scale in order to get an infinitely more rapid coding and processing of information.

The unit in classical computer science is the *bit*, which follows a binary logic. It takes only two values denoted as 0 or 1. On the other hand, a *quantum bit* is associated with the quantum state of a quantum dot assimilated with a system with two levels characterized by the states $|0\rangle$ and $|1\rangle$. Any linear superposition $|\Phi\rangle$ of these two states is also a state materializing the quantum bit with:

$$|\Phi\rangle = \alpha|0\rangle + \beta|1\rangle,\ |\alpha|^2 + |\beta|^2 = 1 \qquad [A2.1]$$

The existence of superpositions of type [A2.1] makes it possible to develop *quantum computers* that are far more rapid than a classical computer.

A.2.2. Quantum dots emergence

Many quantum dots are generated by interface defects of the *quantum wires* such as those engraved in V. This is particularly the case of the quantum wires of gallium arsenide (GaAs) that are used as quantum wells, and of the alloy of gallium arsenide and aluminum (GaAlAs), which operates as a potential barrier. These two materials are deposited by *vapor phase epitaxy* on a GaAs substrate. V-grooves have previously been engraved on this substrate by *photolithography* (*set of operations enabling the transfer of an image on a substrate*), spaced at 4 µm and with a depth of 1 µm. Growth takes place in the following order [END 07]:

1) GaAs layer serves as buffer;

2) barrier layer of $GaAl_{0.6}As$ and then $GaAl_{0.3}As$;

3) 5 nm layer of GaAs well material;

4) another barrier of $GaAl_{0.6}As$ and then $GaAl_{0.3}As$.

The electron microscopy image of the stacking is shown in Figure A2.1. The studied sample has a surface of 2 mm × 2 mm, which represents 500 wires.

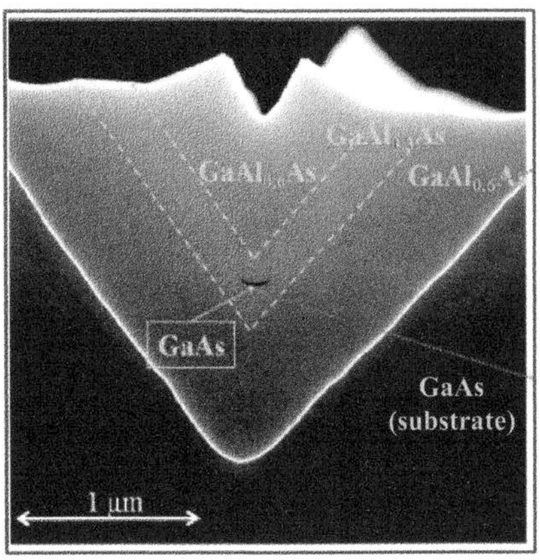

Figure A2.1. *Electron microscopy image of a cross-section of a GaAs quantum wire [END 07]*

Since the axis of the wire is *Ox*, the confinement is realized along the two directions *Oy* and *Oz*. It is, nevertheless, worth noting that the origin of the confinement is different in the two directions *Oy* and *Oz* perpendicular to the wire. In the growth direction (*Oy*), it is the gap difference (600 meV) between the two materials that allows the confinement of carriers in the material whose gap is the lowest (here GaAs). On the other hand, in the *Oz* direction, confinement is due to the fact that the GaAs film is thicker at the bottom of the V-groove.

A2.3. Confinement energy

The energy of an electron (a hole), calculated at the lowest level of the band, in the conduction band (valence band) is given by the sum of a confinement energy

E^{conf} and a band energy E^{free}. Knowing that m_e* (m_h*) represents the effective mass of the particle in the conduction (valence) band and L_y and L_z are the confinement transversal dimensions, we have:

$$E_{e(h)} = E_{e(h)}^{conf} + E_{e(h)}^{free} = \frac{\hbar^2\pi^2}{2m_{e(h)}^*}\left(\frac{n_y^2}{L_y^2}+\frac{n_z^2}{L_z^2}\right)+\frac{\hbar^2 k_x^2}{2m_{e(h)}^*} \quad \text{[A2.2]}$$

Expression [A2.2] shows that the energy in the GaAs quantum wire is only partially discretized, since the conduction electrons can freely move along the free direction Ox of the wire. Confining the carriers in the third direction of space (free axis x of the wire), we obtain a quantum dot. Because real structures are not perfect, having various defects and inhomogeneities, this confinement naturally emerges when quantum dots are maintained at a temperature below 10 K.

To obtain this, the sample is immersed in a *cryostat* that enables the operation at 10 K. It is worth noting that the cryostat comprises a tank filled with liquid nitrogen at 77 K surrounding a tank with liquid helium at 4.2 K, the two tanks being separated by a vacuum enclosure that limits thermal diffusion between the two tanks. The sample is fixed on a sample holder that is itself in contact with the cold finger containing liquid helium. This is how the sample is maintained at a temperature of 10 K during the experiment. When 0D confinement is realized, the energy of an electron (e) (respectively of a hole (h)) in a quantum dot is therefore uniquely a confinement energy given by the following expression [END 07]:

$$E_{e(h)}^{conf} = \frac{\hbar^2\pi^2}{2m_{e(h)}^*}\left(\frac{n_y^2}{L_y^2}+\frac{n_z^2}{L_z^2}+\frac{n_x^2}{L_x^2}\right) \quad \text{[A2.3]}$$

In [A2.3], the quantum numbers n_x, n_y and n_z are strictly positive. The lowest confinement energy is given by the triplet (n_y, n_z, n_x) = (1, 1, 1). Moreover, the energy [A2.3] is similar to expression [1.133] corresponding to the confinement energy in a cubic quantum dot of edge a. In the present case, the quantum dot constituted of GaAs-based quantum wires is parallelepipedic. The transversal dimensions of confinement are L_y = 5 nm, L_z = 15 nm and L_x = 50 nm.

Appendix 3

Transparency of a Potential Barrier, Resonance

A3.1. Expression of transparency

Physical phenomena related to the transparency of a *rectangular potential barrier* of height V_0 and width a have been studied in Chapter 1. We have studied the particular case of a particle of total energy $E < V_0$ that moves toward the barrier from a point of abscissa $x < 0$. The profile of the barrier is schematically represented in Figure 1.7. This appendix completes the description of the behavior of the particle in contact with the barrier and analyzes Schrödinger's equation in the three zones I, II and III for $E > V_0$.

If Schrödinger's stationary equation is applied to zones I, II and III, expressions [1.102], [1.103] and [1.104] are obtained as follows:

$$\frac{d^2\Phi_{\text{I}}(x)}{dx^2} + \frac{2mE}{\hbar^2}\Phi_{\text{I}}(x) = 0\,;\ \frac{d^2\Phi_{\text{II}}(x)}{dx^2} + \frac{2m}{\hbar^2}(E - V_0)\Phi_{\text{II}}(x) = 0 \qquad \text{[A3.1]}$$

$$\frac{d^2\Phi_{\text{III}}(x)}{dx^2} + \frac{2mE}{\hbar^2}\Phi_{\text{III}}(x) = 0$$

We put:

$$k_{\text{I}}^2 = k_{\text{III}}^2 = \frac{2mE}{\hbar^2}\,;\ k_{\text{II}}^2 = \frac{2m(E - V_0)}{\hbar} \qquad \text{[A3.2]}$$

Solutions to equations [A3.1] can be written as follows:

$$\Phi_{\mathrm{I}}(x) = Ae^{ikx} + Be^{-ikx}\,;\ \Phi_{\mathrm{II}}(x) = Ce^{ik'x} + De^{-ik'x}$$

$$\Phi_{\mathrm{III}}(x) = Fe^{ikx} \qquad [A3.3]$$

Let us now express the boundary conditions in $x = 0$, and then in $x = a$. Using [A3.3], we get:

$$\begin{cases} x = 0 : A + B = C + D \\ x = a : Ce^{ik_{\mathrm{II}}a} + De^{-ik_{\mathrm{II}}a} = Fe^{ik_{\mathrm{I}}a} \end{cases} \qquad [A3.4]$$

$$\begin{cases} x = 0 : k_{\mathrm{I}}(A - B) = k_{\mathrm{II}}(C - D) \\ x = a : k_{\mathrm{II}}(Ce^{ik_{\mathrm{II}}a} - De^{-ik_{\mathrm{II}}a}) = k_{\mathrm{I}}Fe^{ik_{\mathrm{I}}a} \end{cases} \qquad [A3.5]$$

The transparency of the barrier is $T = |F|^2/|A|^2$. Let us eliminate the coefficients C and D expressing them as a function of F. Using [A3.4] and [A3.5], we get:

$$\begin{cases} 2C = \dfrac{k_{\mathrm{I}}}{k_{\mathrm{II}}}(A - B) + (A + B) \\ 2D = -\dfrac{k_{\mathrm{I}}}{k_{\mathrm{II}}}(A - B) + (A + B) \end{cases} \qquad [A3.6]$$

$$\begin{cases} Ce^{ik_{\mathrm{II}}a} + De^{-ik_{\mathrm{II}}a} = Fe^{ik_{\mathrm{I}}a} \\ Ce^{ik_{\mathrm{II}}a} - De^{-ik_{\mathrm{II}}a} = \dfrac{k_{\mathrm{I}}}{k_{\mathrm{II}}}Fe^{ik_{\mathrm{I}}a} \end{cases} \qquad [A3.7]$$

Using [A3.7], we have:

$$\begin{cases} 2Ce^{ik_{\mathrm{II}}a} = \left(1 + \dfrac{k_{\mathrm{I}}}{k_{\mathrm{II}}}\right)Fe^{ik_{\mathrm{I}}a} \\ 2De^{-ik_{\mathrm{II}}a} = \left(1 - \dfrac{k_{\mathrm{I}}}{k_{\mathrm{II}}}\right)Fe^{ik_{\mathrm{I}}a} \end{cases} \qquad [A3.8]$$

Let us transform equations [A3.8] as follows:

$$\begin{cases} 2C = \left(\dfrac{k_{II} + k_I}{k_{II}} \right) Fe^{ik_I a} \times (\cos k_{II} a - i \sin k_{II} a) \\ 2D = \left(\dfrac{k_{II} - k_I}{k_{II}} \right) Fe^{ik_I a} \times (\cos k_{II} a + i \sin k_{II} a) \end{cases} \quad [A3.9]$$

Taking systems [A3.6] and [A3.9] into account, we get:

$$\begin{cases} \dfrac{k_I}{k_{II}}(A - B) + (A + B) = \left(\dfrac{k_{II} + k_I}{k_{II}} \right) Fe^{ik_I a} \times (\cos k_{II} a - i \sin k_{II} a) \\ -\dfrac{k_I}{k_{II}}(A - B) + (A + B) = \left(\dfrac{k_{II} - k_I}{k_{II}} \right) Fe^{ik_I a} \times (\cos k_{II} a + i \sin k_{II} a) \end{cases} \quad [A3.10]$$

Using [A3.10], we express the quantities $(A+B)$ and $(A - B)$ as follows:

$$\begin{cases} (A + B) = Fe^{ik_I a} \times \left(\cos k_{II} a - i \dfrac{k_I}{k_{II}} \sin k_{II} a \right) \\ (A - B) = Fe^{ik_I a} \times \left(\cos k_{II} a - i \dfrac{k_{II}}{k_I} \sin k_{II} a \right) \end{cases} \quad [A3.11]$$

It is now easy to express A and B as functions of F and to deduce the expression of the transparency T of the barrier. Then using [A3.11], we get:

$$\begin{cases} A = \left(\cos k_{II} a - i \dfrac{k_I^2 + k_{II}^2}{2k_I k_{II}} \sin k_{II} a \right) \times Fe^{ik_I a} \\ B = i \dfrac{k_{II}^2 - k_I^2}{2k_I k_{II}} \sin k_{II} a \times Fe^{ik_I a} \end{cases} \quad [A3.12]$$

Using [A3.12], transparency T of the barrier is:

$$T = \left| \frac{F}{A} \right|^2 = \frac{4k_I^2 k_{II}^2}{4k_I^2 k_{II}^2 + (k_I^2 - k_{II}^2)^2 \sin^2 k_{II} a} \quad [A3.13]$$

In order to verify the law of conservation of mass, let us express the probability of reflection R. According to [A3.12], we have:

$$R = \left|\frac{B}{A}\right|^2 = \frac{(k_I^2 - k_{II}^2)^2 \sin^2 k_{II}a}{4k_I^2 k_{II}^2 + (k_I^2 - k_{II}^2)^2 \sin^2 k_{II}a} \quad [A3.14]$$

Summing [A3.13] and [A3.14], it can actually be verified that $T + R = 1$.

A3.2. Resonance

Let us express the transparency T as a function of E and V_0 inserting [A3.2] in expression [A3.2]. We get:

$$T = \frac{4E(E - V_0)}{4E(E - V_0) + V_0^2 \sin^2\left[\frac{\sqrt{2m(E - V_0)}}{\hbar} a\right]} \quad [A3.15]$$

The denominator of equation [A3.15] shows that there are values of the width a of the barrier for which transparency is maximal, therefore $T = 1$. These values have been obtained for:

$$\sin^2\left[\frac{\sqrt{2m(E - V_0)}}{\hbar} a\right] = \sin^2 k_{II}a = 0 \Rightarrow k_{II}a = n\pi \quad [A3.16]$$

Let us take into account the wavelength in zone II. According to [A3.16], we get:

$$k_{II} = \frac{2\pi}{\lambda_{II}} \Rightarrow a = n\frac{\lambda_{II}}{2} \quad [A3.17]$$

Fixing E and V_0, the representative curve of the transparency variations as a function of the width a of the barrier shows that T oscillates periodically between its minimal value $T_{min} = 4E(E - V_0)/[4E(E - V_0) + V_0^2]$ and its maximal value $T_{max} = 1$ as shown in Figure A3.1.

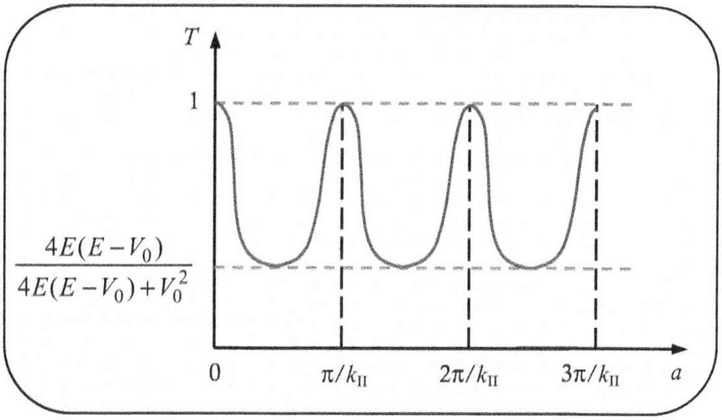

Figure A3.1. *Variations of transparency T of a potential barrier as a function of the width a of the barrier*

Therefore, in zone II a *resonance phenomenon* occurs each time the width a is equal to an integer number of half wavelengths in zone II [A3.17].

Reflected waves undergo constructive interference. For this reason, the *resonance condition* $k_{II}a = n\pi$ corresponds to the values of the width a for which a system of stationary waves can be established in zone II. On the other hand, far from resonances, the waves reflected at the points of discontinuity of the potential undergo destructive interference. The values of the wave function become weak.

References

[ANN 74] ANNEQUIN R., BOUTIGNY J., *Électricité* 2, Vuibert, Paris, 1974.

[ASL 08] ASLANGUL C., *Mécanique quantique 2, développements et applications à basse énergie*, Éditions De Boeck Université, Brussels, 2008.

[ATT 05] ATTAOURTI Y., *Mécanique quantique, une approche analytique*, Tome I, Afrique Orient, Casablanca, 2005.

[BAR 06] BARISIEN T., "Fil quantique idéal", available at: www.cnrs.fr/publications/images delaphysique/couv.../15_ Fil_quantique_ ideal.pdf, 2006.

[BAS 17] BASDEVANT J.-L., *Introduction à la mécanique quantique*, 2nd ed., Éditions De Boeck Université, Brussels, 2017.

[BAY 17] BAYE D., DUFOUR M., FUKS B., *Mécanique quantique, une introduction générale illustrée par des exercices résolus*, Ellipses Éditions Marketing S.A., Paris, 2017.

[BEL 03] BELORIZKY E., *Initiation à la mécanique quantique, approche élémentaire et applications*, Dunod, Paris, 2003.

[BIÉ 06] BIÉMONT E., *Spectroscopie atomique*, Éditions De Boeck, Brussels, 2006.

[BLI 15] BLINDER R., Étude par Résonance Magnétique Nucléaire de nouveaux états quantiques induits sous champ magnétique: condensation de Bose-Einstein dans le composé DTN, Thesis, Université Grenoble Alpes, 2015.

[CHP 78] CHPOLSKI E., *Fondement de la mécanique quantique et structure de l'enveloppe électronique de l'atome*, Éditions Mir, Moscow, 1978.

[COH 77] COHEN-TANOUNDJI C., DIU B., LALOE F., *Mécanique Quantique* 1, Hermann, Paris, 1977.

[COH 92] COHEN-TANOUNDJI C., DIU B., LALOE F., *Mécanique Quantique* 2, Hermann, Paris, 1992.

[DIR 67] DIRAC P.A., *Les principes de la mécanique quantique*, Presses Polytechniques et Universitaires Romandes, Lausanne, French translation of "The Principales of Quantum Mechanics" (4th ed., 1958), 1967.

[DUB 04] DUBIN F., Exciton dans un fil quantique organique, Thesis, Université Paris VI, 2004.

[END 07] ENDERLIN A., Contrôle cohérent des excitations électroniques dans une boîte quantique unique, Report, Institut des NanoSciences de Paris, Paris, 2007.

[FRI 85] FRIDRINE S., MOVNINE S., *Bases physiques de la technique électronique*, Éditions Mir, Moscow, 1985.

[GRI 95] GRIFFITHS D.J., *Introduction to Quantum Mechanics*, Prentice Hall Inc., Upper Saddle River, NJ, 1995.

[GRI 08] GRIBON J.R., *Le chat de Schrödinger, la physique quantique et le réel*, Éditions Alphée, Jean-Paul Bertrand, 2008.

[HLA 00] HLADIK J., CHRYSOS M., *Introduction à la mécanique quantique*, Dunod, Paris, 2000.

[LAH 17] LAHOUAL M., Étude des propriétés optoélectroniques d'une diode laser à puits quantique à base du $Mg_xZn_{1-x}Se$, Thesis, Université Mohamed Khider–Biskra, Algeria, 2017.

[MAR 00] MARCHILDON L., *Mécanique quantique*, Éditions De Boeck, Brussels, 2000.

[MOI 16] MOISAN M., KÉROACK D., STAFFORD L., *Physique atomique et Spectroscopie optique*, EDP Sciences, Grenoble, 2016.

[PHI 03] PHILLIPS A.C., *Introduction to Quantum Mechanics*, John Wiley & Sons, New York, 2003.

[PLO 16] PLOTNITSKY A., *The Principles of Quantum Theory, From Planck's Quanta to the Higgs Boson*, Springer International Publishing, Switzerland, 2016.

[SAK 12] SAKHO I., *Mécanique Quantique 1, exercices corrigés et commentés*, Hermann, Paris, 2012.

[SAK 15] SAKHO I., Cours de propriétés électroniques liées au confinement, Master de Physique des Matériaux, UFR Sciences et Technologies, Université Assane Seck de Ziguinchor, Ziguinchor, 2015.

[SIV 86] SIVOUKHINE D., *Cours de Physique Générale, Tome V: Physique Atomique et Nucléaire*, Éditions Mir, Moscow, 1986.

[STA 08] ST-AMAND A., *Physique des ondes*, Presses de l'Université du Québec, Canada, 2008.

[STÖ 07] STÖCKER H., JUNDT F., GUILLAUME G., *Toute la Physique*, Dunod, Paris, 2007.

[TAR 79] TARRASSOV L., *Bases physiques de l'électronique quantique*, Éditions Mir, Moscow, 1979.

Index

K, L

M

N, O

Other titles from

iSTE

in

Waves

2019

BERTRAND Pierre, DEL SARTO Daniele, GHIZZO Alain
The Vlasov Equation 1: History and General Properties

DAHOO Pierre-Richard, LAKHLIFI Azzedine
Infrared Spectroscopy of Triatomics for Space Observation
(Infrared Spectroscopy Set – Volume 2)

RÉVEILLAC Jean-Michel
Electronic Music Machines: The New Musical Instruments

ROMERO-GARCIA Vicente, HLADKY-HENNION Anne-Christine
Fundamentals and Applications of Acoustic Metamaterials: From Seismic to Radio Frequency
(Metamaterials Applied to Waves Set – Volume 1)

SAKHO Ibrahima
Introduction to Quantum Mechanics 1: Thermal Radiation and Experimental Facts Regarding the Quantization of Matter

2018

SAKHO Ibrahima
Screening Constant by Unit Nuclear Charge Method: Description and Application to the Photoionization of Atomic Systems

2017

DAHOO Pierre-Richard, LAKHLIFI Azzedine
Infrared Spectroscopy of Diatomics for Space Observation (Infrared Spectroscopy Set – Volume 1)

PARET Dominique, HUON Jean-Paul
Secure Connected Objects

PARET Dominqu e, SIBONY Serge
Musical Techniques: Frequencies and Harmony

RÉVEILLAC Jean-Michel
Analog and Digital Sound Processing

STAEBLER Patrick
Human Exposure to Electromagnetic Fields

2016

ANSELMET Fabien, MATTEI Pierre-Olivier
Acoustics, Aeroacoustics and Vibrations

BAUDRAND Henri, TITAOUINE Mohammed, RAVEU Nathalie
The Wave Concept in Electromagnetism and Circuits: Theory and Applications

PARET Dominique
Antennas Designs for NFC Devices

PARET Dominique
Design Constraints for NFC Devices

WIART Joe
Radio-Frequency Human Exposure Assessment

2015

PICART Pascal
New Techniques in Digital Holography

2014

APPRIOU Alain
Uncertainty Theories and Multisensor Data Fusion

JARRY Pierre, BENEAT Jacques N.
RF and Microwave Electromagnetism

LAHEURTE Jean-Marc
UHF RFID Technologies for Identification and Traceability

SAVAUX Vincent, LOUËT Yves
MMSE-based Algorithm for Joint Signal Detection, Channel and Noise Variance Estimation for OFDM Systems

THOMAS Jean-Hugh, YAAKOUBI Nourdin
New Sensors and Processing Chain

TING Michael
Molecular Imaging in Nano MRI

VALIÈRE Jean-Christophe
Acoustic Particle Velocity Measurements using Laser: Principles, Signal Processing and Applications

VANBÉSIEN Olivier, CENTENO Emmanuel
Dispersion Engineering for Integrated Nanophotonics

2013

BENMAMMAR Badr, AMRAOUI Asma
Radio Resource Allocation and Dynamic Spectrum Access

BOURLIER Christophe, PINEL Nicolas, KUBICKÉ Gildas
Method of Moments for 2D Scattering Problems: Basic Concepts and Applications

GOURE Jean-Pierre
Optics in Instruments: Applications in Biology and Medicine

LAZAROV Andon, KOSTADINOV Todor Pavlov
Bistatic SAR/GISAR/FISAR Theory Algorithms and Program Implementation

LHEURETTE Eric
Metamaterials and Wave Control

PINEL Nicolas, BOURLIER Christophe
Electromagnetic Wave Scattering from Random Rough Surfaces: Asymptotic Models

SHINOHARA Naoki
Wireless Power Transfer via Radiowaves

TERRE Michel, PISCHELLA Mylène, VIVIER Emmanuelle
Wireless Telecommunication Systems

2012

LALAUZE René
Chemical Sensors and Biosensors

LE MENN Marc
Instrumentation and Metrology in Oceanography

LI Jun-chang, PICART Pascal
Digital Holography

2011

BECHERRAWY Tamer
Mechanical and Electromagnetic Vibrations and Waves

BESNIER Philippe, DÉMOULIN Bernard
Electromagnetic Reverberation Chambers

GOURE Jean-Pierre
Optics in Instruments

GROUS Ammar
Applied Metrology for Manufacturing Engineering

LE CHEVALIER François, LESSELIER Dominique, STARAJ Robert
Non-standard Antennas

2010

BEGAUD Xavier
Ultra Wide Band Antennas

MARAGE Jean-Paul, MORI Yvon
Sonar and Underwater Acoustics

2009

BOUDRIOUA Azzedine
Photonic Waveguides

BRUNEAU Michel, POTEL Catherine
Materials and Acoustics Handbook

DE FORNEL Frédérique, FAVENNEC Pierre-Noël
Measurements using Optic and RF Waves

FRENCH COLLEGE OF METROLOGY
Transverse Disciplines in Metrology

2008

FILIPPI Paul J.T.
Vibrations and Acoustic Radiation of Thin Structures

LALAUZE René
Physical Chemistry of Solid-Gas Interfaces

2007

KUNDU Tribikram
Advanced Ultrasonic Methods for Material and Structure Inspection

Printed and bound by CPI Group (UK) Ltd, Croydon, CR0 4YY

15/09/2026

14962352-0002